Criteria for Energy Storage R&D

A report by the
COMMITTEE ON ADVANCED ENERGY STORAGE SYSTEMS
Energy Engineering Board
Assembly of Engineering
National Research Council. *Committee on Advanced Energy*
" *Storage Systems.*

NATIONAL ACADEMY OF SCIENCES
Washington, D.C. 1976

This study was supported under Contract No. E(11-1)-2587
between the Energy Research and Development Administration
and the National Academy of Sciences.

Library of Congress Catalog Card No. 76-47080

International Standard Book No. 0-309-02530-3

Available from

Printing and Publishing Office, National Academy of Sciences
2101 Constitution Avenue, N.W., Washington, D.C. 20418

Printed in the United States of America

80 79 78 77 76   10 9 8 7 6 5 4 3 2 1

PREFACE

The Energy Research and Development Administration (ERDA)
requested in 1975 that the National Research Council (NRC)
conduct a study on the potential of advanced energy storage
systems. The NRC responded by creating a Committee on Advanced Energy Storage Systems within the Energy Engineering
Board of the Assembly of Engineering. The members of the
committee were selected to provide a representative balance
of potential users, suppliers, and technologists of energy
storage.

The committee held meetings in June and September 1975
to receive information on the ongoing programs and view-
points relating to energy storage, provided by ERDA, the
National Science Foundation, and the Electric Power Research Institute (EPRI), and to organize its own study
plan and future activities.

The ERDA request called for: (a) a comprehensive,
systematic evaluation of the types of energy storage systems applicable to the electric utility, industrial,
residential/commercial, and transportation sectors of the
economy; (b) the formulation of trade-off criteria for
assessing the probable merits of each storage system;
(c) the identification of research and development (R&D)
needed to bring promising systems into wide usage;
(d) strategies to determine the extent and priority of
federal and private sector R&D needed to advance these
systems; and (e) strategies to maximize the application
of the R&D by the private sector.

The committee and the ERDA representatives agreed at
their first meeting in June 1975 that the above goals con-
stituted a broad framework for long-term committee activities and that more specific objectives should be selected
for the immediate study. Both the committee and ERDA concluded that the initial efforts, which are the subject of

iii

this report, would focus on an in-depth analysis of the criteria (in terms of performance characteristics) that advanced energy storage systems would have to meet in order to command a substantial prospect of commercial use. The characteristics should include, where applicable, the following:

- Storage capacity
- Charge/discharge rates
- Replacement lifetime
- Weight, volume, or other physical limits
- Critical safety parameters
- Environmental standards
- Acceptable capital and operating costs

Recognizing the wide range of potential storage applications and performance criteria, the committee divided into six panels. Four of these corresponded to the standard energy-user sectors--utilities, residential/commercial, industrial, and transportation. In addition, solar-electric and fusion power-generating systems were singled out for detailed attention, although when developed these probably will be operated as parts of utility systems.

Each panel had at least two committee members, with one presiding as chairman, and included experts appointed by the committee chairman at the recommendation of the panel chairman. A complete roster of the panelists, together with the list of participants invited for specific topics, is included in Appendix A.

Each panel developed its own approach to the analysis of the energy storage characteristics deemed important in its category, working from guidelines suggested by the full committee. After one to three working sessions each, the panels reported their findings to the full committee in February 1976. The committee then reviewed these findings and provided the topic discussions, which appear in chapters 2 through 7. The committee developed the criteria for the selection of exploratory R&D, which are found in chapter 8.

The decision to focus on the requirements of the storage applications rather than on the capabilities of the storage technologies leaves further work to be done. No attempt has been made to catalog the prospective energy storage technologies or to evaluate their comparative potentials for use. Such evaluations, including an assessment of development priorities for advanced storage concepts, will be the subject of future study by the committee.

COMMITTEE ON ADVANCED ENERGY
STORAGE SYSTEMS

W. KENNETH DAVIS, *Chairman,* Vice President, Bechtel Power
Corporation
PAUL F. CHENEA, Vice President, Research Laboratories,
General Motors Corporation
EDWARD E. DAVID, JR., Executive Vice President--Research
Development and Planning, Gould Incorporated
GERALD L. DECKER, Utilities Manager, Petroleum Production
and Services, Dow Chemical
HAROLD B. FINGER, Manager, Center for Energy Systems,
General Electric Company
KENNETH C. HOFFMAN, Head, National Center for Analysis of
Energy Systems, Department of Applied Science, Brookhaven
National Laboratory
ROBERT A. HUGGINS, Director, Center for Materials Research;
Professor, Department of Materials Science and Engineer-
ing, Stanford University
FRITZ R. KALHAMMER, Department Director, Energy Management
and Utilization Technology, Electric Power Research In-
stitute
HEINZ G. PFEIFFER, Manager of Technology and Energy Assess-
ment, Pennsylvania Power and Light Company
WILLIAM E. SIRI, Energy and Environment Division, Lawrence
Berkeley Laboratory, University of California
RONALD SMELT, Vice President and Chief Scientist, Lockheed
Aircraft Corporation
DAVID C. WHITE, Director, Energy Laboratory and Ford Pro-
fessor of Engineering, Massachusetts Institute of Tech-
nology

DeMARQUIS D. WYATT, *Executive Secretary,* National Research
Council

CONTENTS

SUMMARY

Energy storage systems have an enormous potential for
more effective use of energy conversion equipment and for
facilitating large-scale fuel substitutions in the U.S.
economy. Such storage is complex and cannot be evaluated
properly without a detailed understanding of energy supplies and end-use considerations. In general, a coordinated set of actions will have to be taken in several
sectors of the energy system for the maximum potential
benefits of storage to be realized.

The performance criteria prepared for this report may
be helpful in determining whether prospective advanced
energy systems will have the performance characteristics
that make them useful and attractive and, therefore, worth
pursuing through the advanced development and demonstration stages. The merits of potential systems need to be
measured, however, in terms of the conditions that may be
expected to exist after the R&D is completed. Care should
be taken not to apply too narrow a range of forecasts for
those conditions.

Care also will have to be taken to evaluate specific
storage system concepts in terms that account for their
full potential impact. The versatility of some technologies for use in a number of applications areas should be
included in such assessments.

The performance criteria are not appropriate for evaluating exploratory research proposals. Screening these
against ultimate application criteria would be premature
and, perhaps, counterproductive.

In this study, the committee examined the opportunities
and requirements for storage systems in six current and
prospective application areas and issued the following
findings.

1

ELECTRIC UTILITIES

Utilities provide a large potential market for advanced storage systems to improve the capital effectiveness of their generating equipment and to reduce the use of oil- or gas-fired generating units. Suitable storage systems, located at central generating points, at dispersed locations on the transmission and distribution networks, or at customer sites, could displace as much as 10 percent of the installed primary generating equipment considered to be necessary by 1985-90.

Storage capacities ranging from 15 to several thousand megawatts and other correspondingly broad technical characteristics will provide great latitude for technological innovation. Economic characteristics, however, will have to be much narrower. The revenues required to cover the cost of storage systems will have to be less, in general, than the costs for comparable primary generating equipment.

R&D programs to clarify some of the operational uncertainties, system interactions, and secondary benefits of utility storage systems are indicated, in addition to development and demonstration of promising storage concepts.

RESIDENTIAL/COMMERCIAL APPLICATIONS

On-site thermal storage systems will not reduce energy consumption, but will permit a phasing of energy input relative to space conditioning or water heating requirements. Such storage will be imperative for independent solar installations and will be highly desirable for all-electric or solar-electric systems where an incentive exists to reduce utility demand peaks or to foster fuel substitutions.

The economic appeal of thermal storage systems will depend upon the perception by builders and owners of benefits from reduced off-peak electrical rates to offset fixed costs of storage systems. Improved electric heat pumps or advanced design and construction energy-conservation practices that reduce total energy requirements will also diminish the economic incentives for storage. Therefore, the earliest and greatest overall impacts of residential and commercial storage will probably come from the development of concepts compatible with and beneficial to existing structures. R&D on such systems, however, should be conducted in the context of other cost-effective conservation retrofit practices.

A well-designed research, development, and demonstration program for thermal energy storage systems is warranted to establish cost break-even parameters. In
addition, studies should be carried out concurrently on
the kinds and levels of incentives or other actions that
might foster the use of such systems to achieve national
energy goals, if consumer cost savings do not seem to be
attainable.

INDUSTRIAL APPLICATIONS

The exact form and role that industrial storage could
take is not clear, because industry is so diverse and
complex in the ways in which it uses energy. In many
cases the capability exists for inherent storage concepts
in which the equipment, the processes, or the product itself may be useful as short- or long-term energy accumu-
lators. In other cases, add-on storage devices,
particularly for low-grade waste heat, would be indicated
if the stored energy can be economically transported and
used on or off the site.

Although present energy costs should encourage industry
to improve energy-management practices in which storage
systems could have a potential role, the storage requirements will vary too widely to warrant an R&D program aimed
at the development and demonstration of specific storage
devices. Instead, ERDA should conduct a continuing dialogue with a wide range of industry representatives to
determine the significant storage opportunities for both
inherent and add-on systems, conduct specific analytical
studies to quantify and verify potential benefits, arrange for wide dissemination to industry of information
on storage applications and technologies, and support
selected demonstration programs to prove the practical
applicability and to reduce the technological risk of
industrial storage systems.

TRANSPORTATION APPLICATIONS

The U.S. market for cars using stored, externally
generated energy rather than stored fuel will depend largely upon the intended vehicle use. For short-range, multistop commercial applications over defined routes, electric
vehicles may produce attractive life-cycle cost savings
for fleet owners. By contrast, a sizeable penetration of

the personal market for this type of car will probably await the development of vehicle characteristics, including first cost, that are competitive with those of gasoline-powered cars. Limited-use fleet vehicles using storage batteries are possible with current technologies. An urban shopper-commuter vehicle with a range of 50 miles might be technologically feasible in about 10 yr. The committee concluded, however, that a competitive, electric, family vehicle capable of carrying four to six passengers and having a 200-mile range, 55-mph speed, and a competitive first cost will require at least 25 yr of R&D.

Batteries weighing no more than about 25 percent of the gross vehicle weight will be necessary for competitive performance of family cars. This will require specific energy densities of from 10 to 15 times the level of current lead-acid batteries. At the same time, first-cost comparability will require that these advanced batteries have a specific energy cost considerably lower than that of the millions of automobile batteries now made each year. The advanced batteries also will need to withstand 500 or more deep (80 percent) discharges and 7- to 10-h recharges.

A vehicle combining storage with a conventional gasoline engine might achieve earlier market acceptance. For example, a flywheel, coupled with a small engine, might preserve the range and flexibility of the gasoline-powered vehicle while markedly improving mileage and reducing emissions. Such flywheel concepts need further development to establish their performance, safety, and cost characteristics.

SOLAR-ELECTRIC SYSTEMS

The use of storage systems with solar-electric power systems will generally be required because of fluctuations in solar input energy. Several classes of storage may be needed for a single installation, depending on the type, scale, and interconnections of the solar plant with conventional utility systems. These could range from quick-response units to accommodate momentary input interruptions to large-capacity units that provide power at night or during extended low-insolation periods. Even in those instances where overall storage requirements might be similar to those of conventional utilities, the variable and to some degree unpredictable input may lead to a

preference for storage systems capable of accommodating
widely variable charge and discharge rates in contrast
to fixed-capacity electromechanical devices. In addition,
a higher value will likely be placed on devices that can
undergo frequent cycling from charge to discharge modes.

Isolated cost targets for integral storage elements
of total solar-electric power systems cannot be specified.
Since the comparatively high solar plant costs may be the
greatest obstacle to their widespread deployment, however,
the storage component costs will have to be kept as low
as possible.

FUSION REACTORS

While fusion reactors are not yet technically feasible,
the concepts for full-scale designs of such reactors suggest that successful fusion plants will need large internal
energy storage systems with transfer rates exceeding any
existing today. For fusion plants in the output range of
500-3,000 MW, storage capacities may be as great as 500 GJ
(140 MWh), and for some reactor concepts the energy transfer to and from the storage systems must take place in as
little as 0.1 µs. For many of the reactor concepts the
efficiency of the storage system will be critical to the
maintenance of a net energy balance for the plant.

Internal energy storage problems will have to be solved
as part of the fusion reactor designs. Other forms of
external buffering storage of the plant electrical output
may also turn out to be necessary if the actual systems
operate with variable pulses. Clarifying the storage
requirements and finding appropriate technical solutions
will be an important element in future fusion reactor
developments.

1 GENERAL CONSIDERATIONS

Today's industrial civilizations are based upon abundant and reliable energy. To be useful, raw energy forms must be converted into power sources, most commonly through heat release. For example, steam, which is widely used as a heat source in industrial processes, is obtained through a thermal release of fuel energies into water. Electricity, increasingly favored as a power source, is generated predominately with steam-powered turbogenerators, fueled by fossil or nuclear energy.

Power demands in general, whether thermal or electrical, are not steady. Moreover, some thermal and electrical energy sources, such as solar energy, are not steady in supply. In cases where either supply or demand is highly variable, reliable power availability has in the past generally required energy conversion systems large enough to supply the peak-demand requirements. The results are high and partially inefficient capital investments since the systems operate at less than capacity much of the time.

In the future, capital investments might be reduced if load-management techniques can be employed to smooth power demands, on the one hand, or alternatively, if energy storage systems are used to permit smaller power-generating systems to operate at or near peak capacity irrespective of the instantaneous demand for power, by storing the excess converted energy during reduced demand periods for subsequent use in meeting peak-demand requirements.

Although some energy might be lost in the storage process, critical fuel conservation could be attained by utilizing more plentiful but less flexible fuels such as coal and uranium in applications now requiring scarce oil and natural gas. In some cases, storage systems might also enable the waste heat accompanying conversion processes to be used for secondary purposes.

The opportunities for energy storage are not confined to the industries and utilities. Storage at the point of energy consumption, as in residences and commercial buildings, will be essential to the future use of solar heating and cooling systems and may prove important in lessening the peak-demand loads imposed by conventional electrical, space-conditioning systems. In the personal transporation sector, now dominated by gasoline-powered vehicles, adequate storage systems might encourage the use of large numbers of electric vehicles, reducing the demand for petroleum.

The concept of energy storage *per se* is not new.* The marked increases in fuel costs in the last few years, the increasing difficulty in acquiring the large amounts of capital required for power expansions, and the emergence of new storage technologies has, however, led to a recent resurgence of interest in the possibilities for advanced energy storage systems.

To the energy supplier, energy is a commodity whose value is determined by the cost of production and the marketplace demand. For the energy consumer, the value of energy is in its contribution to the production of goods and services or to personal comfort and convenience. In spite of rhetoric about the merits of alternative national energy-production and consumption patterns in the future, it is likely that energy decisions, in general, will continue to be made from an evaluation of alternate energy costs to attain or heighten these perceived values.

In particular, there is no reason to anticipate that decisions to use or not use energy storage systems will be made on any other basis than that of prospective cost savings in the production or use of energy, unless legislative or regulatory constraints are imposed. Thus, in defining the range of performance criteria necessary for the prospective commercialization of storage systems, the conditions for establishing potential economic viability are a major parameter.

---

*The use of flywheels to smooth intermittent power impulses was recognized shortly after the invention of reciprocating engines in the eighteenth century. Special purpose locomotives have been operated with stored, externally supplied steam for about 100 years. Electric cars were early automobile competitors. The first hydroelectric storage installation in the United States was made in 1929.

The relatively low use of storage systems in the past has resulted from the low costs of energy that this nation enjoyed for many years. Capital investments were less productive for storage than for primary production equipment, including additional energy-conversion capacity. Only industries such as the electric utilities, in which fuel costs were a major expense item, made use of storage systems, and even then in limited quantities. Recent increases in fuel costs have raised the investment value of storage systems, however, and should result in a greater number of storage installations in the future. Nevertheless, the future deployment of storage systems will be highly dependent upon an economic trade-off analysis for each potential installation. A brief examination of the reasons for considering the installation of storage systems indicates both the basic trade-off considerations and the complexity of quantifying them in explicit terms.

The basic objectives of storage installations will include one or more of the following: (a) the reduction of energy consumption, (b) the reduction of energy costs, or (c) the substitution of more plentiful for scarcer energy resources. General economic criteria can be defined for each.

If a storage system is installed to collect and hold waste heat otherwise discharged to the environment, and if the energy thus stored is later used in place of added primary energy, a reduction of overall energy consumption will occur. To be economically feasible, the cost of the replaced primary energy will have to exceed the capitalization, maintenance, and operating costs of the storage system. The stored energy can be considered to be free, since it would otherwise be lost.

Most--probably a preponderance--of the storage applications will be designed for load-leveling rather than for waste energy recovery. The energy to be stored will be in a high-grade rather than waste condition, drawn from the conversion-consumption cycle during periods of excess conversion capability and restored to the cycle during periods of high consumption. Such storage systems will not conserve energy, rather they may actually increase use owing to storage-system inefficiencies. At an efficiency of 75 percent, for example, the overall energy consumption for a task supplied through storage would increase by one-third as compared to a direct primary energy supply. The objectives of such systems will obviously not include a reduction in energy consumption, but will be

either to reduce energy costs or to allow the displace-
ment of scarce fuels by more abundant fuels in the energy-
conversion cycle.

Although seemingly incongruous, a number of situations
exist in which the use of storage may decrease energy
costs even though total energy consumption increases.
These arise because of particular balances between fixed
costs associated with initial capital investments and
energy-variable costs, such as fuel and operating and
maintenance (O&M) costs. Consider, for example, an energy
supplier confronted with a demand for a product that cycles with time, as shown at the top of Figure 1. The supplier will have to invest in enough conversion equipment

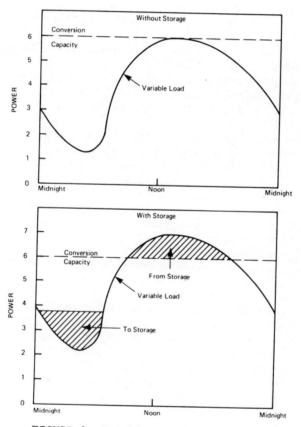

FIGURE 1 Matching load and conversion
capacity, with and without storage.

to meet the peak-demand level, with resultant fixed annual capitalization costs. The energy charges must be structured to enable recovery of capitalization costs (plus output-dependent variable costs) from the cycling load. On the other hand, the cycling nature of the demand load will mean that the installed equipment will have excess capacity part of the time.

If the supplier faces a growing demand for energy, two options will be available for increasing the peak capacity. On the one hand, additional capital investments could be made for added conversion equipment and, to a first order, the same level of costs per unit output could be maintained. As an alternative, the supplier could add a storage system to the present conversion equipment, as shown at the bottom of Figure 1. By so doing, the supplier could increase the output during "valleys" in demand, store this excess output, and release it to satisfy the peak demands that are in excess of the current conversion capacity.

The cost of the stored input energy would, to a first order, be only the variable conversion costs, since the capitalization costs would already be covered by the cycling load charges. The output energy costs would be higher by the reciprocal of the storage efficiency and by the sum of the O&M and capitalization costs of the storage equipment itself. If these storage system costs are lower than the total costs for added primary conversion, then the cost of the added energy output from storage would be less than the cost of equivalent energy from expanding conversion capacity.

Not only might the supplier benefit from storage systems, but under certain circumstances so might the user (in addition to the potential savings from reduced production costs). The user may be required to pay a premium rate for energy during peak-demand periods. Differential charges are common for industrial customers of electric utilities, who are under increasing pressure to apply such charges to commercial and residential customers as well.

Under such circumstances, the user might elect to store cheaper off-peak energy for consumption when higher on-peak rates prevail. The economic viability of user storage systems would depend upon the costs of capitalization, O&M, and increased energy usage due to storage inefficiency being less than the energy savings afforded by the rate differential.

A third major objective of advanced energy storage systems is the conservation of scarce fuels by substituting more abundant ones. At the consumer level, electric vehicles powered by storage batteries charged from nuclear or coal-fueled generating stations might replace conventional gasoline-powered automobiles. At the supplier level, storage might enable excess off-peak nuclear-powered generating capacity to supply peak-power demands in lieu of gas-fired turbogenerators.

The economic trade-offs in fuel substitution decisions will vary with the storage application and with externally imposed constraints. In the transportation sector, for example, the economic motivations of individual buyers are unpredictable. First-cost considerations have some market impact. Other decisions are influenced by relative operating economy (as measured in miles per gallon). Only a few decisions, particularly those of fleet owners, seem to be based on life-cycle costs. The economics of car buying will be particularly complex if gasoline- and storage-powered vehicles are marketed in competiton with each other. Comparisons of first, operating, and life-cycle costs will be given varying degrees of weight in individual decisions, but performance, range, and operational flexibility may continue to be of equal or greater importance.

At the more sophisticated energy-supplier level, the economic bases for making storage decisions for fuel-substitution purposes can be more easily projected. In the first place, unless external constraints are applied, storage decisions will only be made if net cost savings can be projected. The economic considerations will be the same as those discussed previously, but with the added possibilities and returns from displacing specific fuels from expansion considerations.

If the energy supplier is prohibited by law or regulation from installing conversion equipment using specific scarce fuels without regard to cost, economic considerations will obviously be removed from the fuel-substitution question. The supplier will still need to determine whether or not storage will be cost-effective for the more limited range of available alternatives.

Thus, the basic economic criteria that will be applied in making storage-system evaluations are not obscure (except, perhaps, for those involving personal transportation vehicles). The development of quantitative guidelines that will be useful in forming *a priori* judgments on the potential economic viability of specific systems that still have to undergo R&D is, however, extremely complex. In a large measure, this is a consequence of future uncertainties.

The ultimate decision to proceed with the actual installation of a storage system will depend on comparative technical/economic analyses in which, as has been noted, the primary economic variables are capitalization, energy input, and O&M costs. These costs must compare favorably with those of alternative approaches. Since, with rare exceptions, the useful life of a storage system will be many years, the analyses must project fuel, labor, and materials costs. Such projections are uncertain, at best, in the near term. They are even more fraught with uncertainty when the time required for R&D is added.

In developing economic guidelines for potential storage applications, the committee avoided specifying future economic scenarios, while fully realizing that such scenarios are a necessary feature in ultimate decision making. Instead, the committee has emphasized the nature of the economic trade-offs, illustrated with appropriate examples, that will have to be favorably met for commercialization.

The economic complexities that led the committee to favor the generalized approach to the identification of economic criteria suggests, as a corollary, that the potential economic viability of energy storage systems proposed for advanced development and demonstration be assessed for a range of potential economic scenarios. Rejection of concepts as lacking economic feasibility under too narrow a projection of future possibilities could result in a dearth of useful alternatives if events take a markedly different course.

While there is little merit in supporting development and demonstration projects that will obviously fail to meet marketplace requirements, it may be prudent to pursue the technical validation of significant concepts without placing undue weight on any favored economic outlook. (The aftermaths of the 1973 oil embargo amply illustrate the fallibility of energy price predictions.) The full R&D cycle for advanced energy storage concepts may, in many cases, cover a minimum of 7-10 yr. Implementation of decisions to deploy storage systems may require another 2-10 yr, depending on the scale and complexity of the system. Because of the economic perturbations that may occur over such lengthy time intervals, the national interests of the future may well warrant the conduct of a sound technical, though currently economically questionable, R&D program as insurance in the event of such perturbations.

OPPORTUNITIES

The electric-utility industry offers a potentially large
and diverse market for advanced energy storage systems.
The utility motivations may be grouped into three general
classes: economic, environmental, and quality-of-service
considerations. These may operate independently or rein-
force each other.

The utility industry may be able to reduce its capital
requirements for future service expansions and reduce fuel
costs of power generation by installing storage systems.
A general case can be made for storage systems whenever
the annual income required to cover capital and operating
costs is less for a storage installation than that required for primary generating equipment supplying the
same service loads and periods or for methods of load control or other means for accomplishing the same purpose.

The environmental impact of power systems has moved to
the forefront of public concern in recent years. Environmental evaluations of existing and new technologies have
become a major factor in their application on power systems. Several of the storage-system concepts now under
consideration may reduce the environmental impact of
utility expansions as compared with primary generating
equipment. Thus, storage systems could not only reduce
costs, but might also reduce the time period required for
approval and implementation of capacity additions.

The quality-of-service motivations can range from the
use of storage systems to minimize load fluctuations and
thereby reduce maintenance outages on base-load generating
equipment to the protection of future systems against
the potential nonavailability of gas or liquid fuels for
peak-load generation.

Utility storage systems are intended to receive electrical energy generated during off-peak hours, to store
this energy by mechanical, thermal, chemical, or electrical means, and to release the stored energy (less storage
losses) during periods of high demand. When charged from
conventional primary generation equipment, storage provides a method of meeting peak loads with energy generated
by base-load units. Figure 2 shows the typical impact
that storage might have on utility operations. The upper
part indicates the weekly variation in power demand and
the usual apportionment of base, intermediate, and peaking capacity to satisfy the demand. The lower part illustrates how, with an increase in base-load capacity and
the addition of energy storage, the demand load could be
met without the use of peaking generating equipment.

Although often conceptualized in terms of peaking capacity, storage systems could be economic alternatives to
either peaking or intermediate generation expansions.
Intermediate generating units frequently have capacity

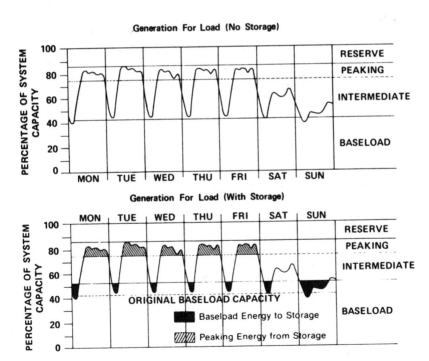

FIGURE 2 Typical weekly load curve of an electric utility.

factors of 25 percent or more as compared to 10 to 15 percent for peaking units. If the intermediate units use high-cost fuels as compared to the base units, the fuel savings with storage systems may more than offset the added capital costs associated with the increased storage-system capacity. Depending on the requirements of the particular utility system and the specific storage-system characteristics, storage systems could be economically attractive for either peaking or intermediate generation modes using either daily or weekly storage cycles.

In addition to serving as an alternate secondary mode for providing system capacity, storage systems may create other system benefits that will enhance their values. These added values may or may not be realized in specific utility installations. Their worth is highly dependent on individual utility characteristics, but qualitative benefits can be described as a function of the storage-system deployment within the utility generating/trans-mission/distribution complex.

Central storage systems, installed to have the same transmission interfaces as the primary generating equip-ment, may yield the following additional benefits, depend-ing on the storage-system technology:

- *Fast response to load swings*. In many instances, rapid fluctuations in load are carried through the system to the turbine with consequent operating problems. Stor-age systems capable of fast load response with no negative side effects would have a comparatively higher value to the system.
- *Reduce the swing of base-load units*. Base-load units operate with maximum efficiency at or near maximum output. As plant output is reduced, plant heat rate per kilowatt-hour output increases. A storage system charging at a rate that would keep the base-load unit at maximum efficiency would have an added value corresponding to the fuel savings for the on-line power.
- *Spinning reserves*. Either idling combustion turbines or base-load units operating below design capacity are used by most utilities to insure adequate backup in case a major unit trips off the line. This reserve requirement might be met by storage systems requiring no warm-up time. The fuel savings resulting from the more efficient use of base-load units and the elimination of idling combustion turbines would add to the value of the storage system.
- *Minimum loads*. The output of base-load generating units can only be reduced to about 30 to 40 percent of

the rated power before turbine operating problems are
encountered. In a 24-h period, the demand-load swings
often exceed those limits, therefore it becomes necessary
to shut some units down. Restarting them a few hours later
is both expensive and hard on the unit. Storage systems
with the capacity to maintain a demand load of at least
40 percent of the peak would eliminate that problem.

If technically and economically satisfactory, smaller
storage devices can be developed, they might be installed
close to the ultimate load rather than in a central storage
system. Dispersed in this way over the electric power system, these devices could perform the same system peak-
shaving and load-leveling functions that can be provided
by central energy storage installations. In addition,
dispersed storage would have several unique capabilities
that might add to its value. The most important of these
include:

• *Increased transmission and distribution capability.*
By reducing the load levels experienced in the transmission
and distribution (T&D) networks upstream of the storage device, the line losses would be reduced for given customer
loads; or, conversely, the load capacities would be increased for a given line installation. These factors would
add only a slight value to a dispersed storage system
added to an existing T&D network. If these same factors
were incorporated into long-range planning, however, they
could minimize or defer T&D expansions and add considerably
to the apparent value of the storage systems.
• *Load protection from T&D outages.* The presence of
energy storage near the load could reduce the impact of
T&D outages on load requirements. The associated increase
in system reliability could reduce the reserve generation
margins required to maintain specified levels of system
security.
• *Incremental growth.* If reliable, modular, factory-
built, and easily transportable storage devices were available to utilities, incremental expansions of storage-system
installations could be programmed with both load and T&D
growths. This flexibility could aid utilities to make
rapid adjustments to changing economic and institutional
factors with reduced financial risk and costs.

The value credits that might be accrued from the above
storage benefits are not well quantified. The committee
recommends that studies be undertaken to establish probable
values.

The consideration of storage devices is primarily fixed by: (a) the difference between a utility's average cost of incremental valley energy and the average peak-period cost, (b) the efficiency of the storage device, (c) the first cost of the storage device as compared to the cost of conventional generation, and (d) the cyclical load characteristics experienced by the utility. Since most storage devices will be capacity-limited to daily or weekly storage cycles, as contrasted to seasonal cycles, the load characteristics of interest are those experienced in an approximately 7-day period.

These four considerations, properly weighed, can be used to forecast the probable amount of energy storage that will be deployed by utilities.

The electric-generating mix expected to be available to a utility in the future will influence decisions regarding the installation of storage devices. For example, one utility, after examining the four variables cited, determined that up to 10 percent of its generation capacity should be pumped-hydro storage. However, the utility later changed its plans for future base-load generation by reducing the planned percentage of low-cost nuclear generation. The reduced increment in valley-to-peak period costs in future years then showed that additional storage was no longer justified.

Conversely, the increased future availability of economically feasible storage devices may influence the planned generating mix, giving greater value to generating units having low incremental valley energy costs. Such units will not be installed, of course, unless warranted by expanding electrical base loads.

The installation of storage devices will probably be highly variable through 1985. With the limited technologies available in the near future, many utilities may defer storage installations. Others may install as much as 15 percent of the generating capacity. Most of this will be pumped-hydro storage, which currently offers the most promising combination of efficiency and low first cost.

Overall, pumped storage was about 1.8 percent (8,800 MW) of installed U.S. generation capacity at the end of 1974. This is projected to grow to about 23,400 MW, or 2.8 percent, by the mid-1980's. Other current storage technologies (primarily thermal and underground compressed air) will probably add no more than an additional 20 percent to storage capacity, for a maximum projected storage of about 3.5 percent of installed generating capacity by 1985.

Beyond 1985, the technical and economic development of
advanced storage systems will pace the rate of installations. If technologies can be developed with acceptable
capital costs, 2- to 4-yr construction lead times, and efficiencies at or above the present 60-75 percent levels,
storage devices might displace a large portion of the
future peaking and intermediate generation expansion.
Some utilities that are dependent on oil or gas might justify storage up to 15-20 percent of their capacities.

On solely economic bases, storage might be used for up
to 30 percent of the installed generating facilities, if
no other means for accomplishing the same purposes become
available. However, since storage systems are secondary
forms of generation capacity and are dependent upon base-
load capacity for charging energy, their overall reliability and system performance will be a function of the
reliability and availability of excess off-peak, base-load
capacity. The effects of significant amounts of storage
capacity on a system's generation mix, generation reserve
margin, and reliability of service have not yet been fully
defined, largely because operational data are not widely
available.

If, in the 1985-90 time period, total installed reserves are about 18 percent, it is reasonable to project
that the installation of energy storage systems will not,
in all probability, exceed about 10 percent of the installed generating capacity rather than 30 percent, even
if the improvements sought in advanced energy storage
devices are forthcoming.

PERFORMANCE CRITERIA

The extent to which storage systems can be effectively
used by utilities depends on many factors, including:

- The system load characteristics
- The system generation mix
- The amount of generating capacity available for
charging the storage system
- The duty cycles required of the storage system
- The storage-system performance characteristics (use-
ful life, efficiency, reliability, unit capacities, and
so forth)
- The incremental cost of off-peak energy in compari-
son to incremental peak-energy costs
- Storage-system capital and O&M costs
- The economics of alternative generating capacities.

The Public Service Electric and Gas Company (PSEG) of New Jersey has recently assessed specific storage-system technologies in a study jointly supported by ERDA and EPRI.[1] The study illustrated the complex interactions of the above factors. For the purpose of outlining the general performance characteristics needed in a storage device for utility applications, this discussion centers on three main topics: (a) storage capacities, (b) storage-operating characteristics, and (c) storage economics.

Storage Capacities

The demand-load requirements on utilities vary widely. In general, utilities in the northeastern United States experience winter peaks associated with the heating season, while those in the South have summer peaks arising from air conditioning loads. Some midwestern utilities have both peaks. Not only do the seasonal peaks vary, but the daily variations may be marked. Depending on the customer load mix, the peaks may spike for relatively short intervals once or twice a day (for predominately residential/commercial service), or may rise to nearly constant values for as long as 17 h (for large industrial loads). These variations create different storage-capacity requirements and off-peak charging capabilities.

The maximum amount of on-peak energy that could be furnished from excess off-peak generating capacity is a function of the system load factor and the base-load capacity. The PSEG study indicates that after base-load capacities are corrected for forced and maintenance outages, utilities with annual load factors ranging from 80 percent to 40 percent might have off-peak capabilities to generate from 5 to 13 percent of the annual energy produced for load, respectively. Thus, for a storage system of no losses, a utility with a 60 percent annual load factor could support a maximum of about 10 percent of its total annual requirements from off-peak energy.

Because the distribution of off-peak energy is not uniform over the year, only about 70 percent can actually be stored consistently. In order to reach this practical limit, storage systems would have to operate on a weekly cycle. The PSEG analysis of utility load curves indicates that about 93 percent of the on-peak energy requirements occur on weekdays, as compared to 7 percent on weekends. On the other hand, about 45 percent of the off-peak energy is available on weekends, as contrasted with 55 percent

on weekdays. If storage capacity is limited to daily
cycles, then only about one-half of a utility's off-peak
energy limit can be effectively utilized.

These limitations on the maximum practical amount of
storage capacity that can be supported by electric utilities with 50 and 80 percent annual load factors and 100
percent storage efficiency are summarized in Table 1.

The maximum capacity estimate of 19 percent of peak
load for peaking-duty application is based on the use of
weekend and weekday off-peak energy to charge storage systems with 2- to 12-h discharge capability (in 2-h increments) in steps of approximately 3 percent of peak load.

The above levels of output are predicated on storage
systems that are 100 percent efficient. The levels will
be reduced in direct proportion to lower efficiencies.
If the storage efficiency is only 75 percent, for example,
the weekly cycle capacity levels would drop to 7-14 percent for peaking-duty applications and to 4-7 percent for
intermediate duty.

Unit storage-system capacities will be a function of
the location of the storage device within the utility
system. The peak loads of U.S. utilities vary from less
than 10 MW to almost 17,000 MW. The average is about
1,500 MW. Annual energy produced for load ranges from
27,000 MWh to 93,500,000 MWh with an average of about
7,900,000 MWh.

These ranges suggest that a correspondingly wide range
of central storage capacities might be used. Minimum useful sizes will probably be set by economics of scale,
maximum sizes by technology limits.

Storage at the transmission level in most of the United
States will be tied to a grid with a capability of 10,000
MW or more. At this level, less than 500 MW of storage

TABLE 1 System Storage Limits

| Mode of Energy Storage Operation | System Generation Application | | | |
| | Peaking Duty | | Intermediate Duty | |
	Peak Load (%)	Installed Capacity (%)[a]	Peak Load (%)	Installed Capacity (%)[a]
Daily cycle	7-13	5-10	3-6	2-5
Weekly cycle	10-19	8-15	6-11	5-9

[a]Installed capacity assumed to be 1.25 times peak load.

power capacity would have little impact. A weekly energy
component would be desirable on most systems, making 5,000
MWh a likely minimum energy size.

Storage on the distribution system would have a capacity
corresponding to the peak load at a substation. A typical
size might be 15 MW power with 150 MWh energy, and 50 MW
with 500 MWh would be a maximum. The former size would be
typical for a 69 Kv to 12 Kv substation, the latter on a
220 Kv to 69 Kv substation. Much smaller devices would
have limited use.

The growth of on-site electrical generation by commercial and industrial operations via total-energy systems
is an interesting development that could affect dispersed
storage demand considerably. These systems can use energy
optimally as electricity, as heat, or as direct mechanical
energy. The demand and production of the different energy
forms do not generally coincide, so that storage of one
form or another will be an essential part of the systems.
In industrial operations, in particular, the process energy
use will dominate and the electrical production will be a
dependent variable. The utility would act as a source or
sink for unmatched electrical demand or output, but on-site
storage of electrical energy could optimize the load from
the point of view of both the utility and industry.

The main form of storage at residences will probably
be the storage of thermal energy for space conditioning and
hot water supply. The development of suitable batteries
would make electrical storage feasible, but in most cases
space conditioning and hot water load storage would provide
the utilities with the desired load flattening.

Storage-Operating Characteristics

A major factor in determining the feasibility of storage
is the shape and impact of the daily load curve on the
availability of energy for charging the storage device.
An urban utility with a summer peak may have a peak of
moderate duration during the day and a deep valley of short
duration at night. Thus, the storage device may be required to supply energy for 10 h during the day while significant charging capacity is only available for 6 h at
night. Such a storage device would either have to be
charged at a faster rate than it is discharged or operate
on a weekly cycle with part of the charge taking place
over the weekend.

The diverse operating conditions for utilities will
cause substantial variations in the duty cycles of storage

TABLE 2   Duty-Cycle Characteristics

Duty-Cycle Character- istics	Type of Operation			
	Peaking Duty		Intermediate Duty	
	Daily Cycle	Weekly Cycle	Daily Cycle	Weekly Cycle
Discharge time (h/day)	2-8	2-8	9-14	9-14
Charge time (h/day)				
Weekday	5-9	5-9	5-9	5-9
Weekend	--	14-34	--	14-34
Charge/dis- charge ratio	0.8-2.1	0.1-2.1	1.3-3.7	0.8-2.4
Storage capabil- ity (h)	2-8	4-26	9-14	17-47
Annual operating hours (dis- charge time)	350-1,600	350-1,600	2,300-3,600	2,300-3,600

systems. Table 2 summarizes the typical range of cycle
characteristics for different applications. The values
assume 75 percent storage efficiency.

Storage Economics

The economic justification for storage systems assumes
that the annual income needed to cover capital and operating costs will be less than that required for primary
generating equipment supplying the same service loads and
periods. In general, there will be fuel cost savings as
compared with primary generating equivalents, but often
at the expense of higher initial capital costs. At times,
the capital market is such that the utilities have difficulty raising funds. Such a situation currently exists
and will probably continue for some time. Furthermore,

at present return levels, there is little incentive for
utilities to seek additional capital.

On the other hand, expense dollars can be recovered
rapidly rather than over the life of the plant. Short-
term financing costs that might termporarily dilute earn-
ings could jeopardize a utility's capability to acquire
additional capital in the future. In such circumstances,
utilities might delay or forego the long-term benefits
of storage in favor of other operating solutions.

Future economic uncertainties further complicate the
situation. The construction and installation of the
storage systems may vary from 2 yr for batteries to a
decade for pumped storage. Once systems are operational,
their useful lives may range from 20 to 40 yr. Conse-
quently, the decision to install a storage system must be
based on anticipated system loads, load characteristics,
and generating-capacity mix for a very extended period.
Regional uncertainty about the future economic outlook,
life-style changes, or the availability of low-cost
charging power for the storage system may lead to differing
investment decisions if alternative technical solutions
are feasible.

Uncertainty also stems from the technical dynamics of
alternative solutions to the problems that make storage
systems potentially attractive. Technologies cannot be
frozen at their current state of the art in drawing system comparisons. Decisions must include the possibility
of technical advances that might change the comparisons.
For example, any decision regarding the installation of
storage systems as a hedge against the possible future
unavailability of oil or gas supplies should include an
assessment of the potential for fueling peak-load gas
turbines directly from coal or coal derivatives. Differing assessments of the probability of such developments
might alter decisions markedly.

Other alternatives must also be weighed. Storage systems are intended to provide an alternative to real-time
primary generation for peak loads. Any institutional
infrastructure changes, such as time-of-day and lifeline
rate structures that would tend to minimize or eliminate
the peaks would alter the basis for installation of the
systems. Therefore, the probability of their adoption
and their potential impact on peak loads must be included
in the decision process. Decentralized rate-setting au-
thority means that the assessment and conclusions will
probably differ for each utility. Extensive application
of load-management techniques might also minimize or
eliminate load peaks.

In the light of these uncertainties and the differences
in economic conditions for individual utilities, the
committee cannot lay out generally valid, quantitative
cost and economic criteria for the viability of storage
systems. Each installation will have to be evaluated on
a case-by-case basis against primary generation alternatives, using the utilities' standard analyses methods.
For example, application of system expansion models with
and without storage and determination of break-even costs
can be made on the basis of equal present worth of all
future revenue requirements. This type of analysis was
performed in the PSEG report cited earlier. Discussion
of the PSEG study results is beyond the scope of this
report. However, a discussion of the major economic factors is useful in establishing some approximate cost
criteria for energy storage.

Briefly, revenue requirements consist of three categories: annual fixed carrying charges for equipment capital
costs, energy or fuel charging costs, and operation and
maintenance (O&M) charges. Annual fixed costs must cover
the anticipated rate of return on invested capital, taxes,
depreciation, and a construction compound-interest factor
(CCIF), which accounts for the cost of money during construction. The anticipated rate of return for utilities can
be taken as a nominal 10 percent. Depreciation rates will
vary inversely with the expected life of the installation.
The CCIF depends on the length and expenditure rates during
construction phases. It will vary from about 1.05 for
typical 3-yr construction periods to 1.40 for 8-yr periods.

Typically, the total annual carrying charges will range
from 15 percent of capital cost for storage systems having
25-yr lifetimes and 3-yr construction periods to about
22 percent for systems requiring 8-yr construction periods.
Systems with limited lifetimes, such as current batteries,
might require capital carrying charges as high as 27 percent for 5-yr lifetimes.

In general, capital charges will represent the largest
part of revenue requirements, especially if the storage
system's annual use (load factor) is relatively low, as
in peaking service. As discussed below, the allowable
capital charges can represent a major constraint for
storage technologies.

Charging costs for storage systems are determined
from the average valley excess (nonload) unit costs from
the base load, multiplied by the number of hours of storage operation per year and divided by the round-trip storage efficiency. Depending on the source of base-load

charging power (nuclear or coal), the storage input energy
will probably fall in a range from 5 to 20 mills/KWh.

The O&M costs of storage systems are normally expected
to be substantially less than either the annual fixed
carrying charges or the energy charging costs.

The relative importance of these cost factors and an
indication of the break-even investment for storage systems can be illustrated by a representative example.
Suppose that alternatives are being considered for a peaking primary generator using a gas turbine, costing $130/KW,
burning fuel costing $2.50 per million Btu at a heat rate
of 12,100 Btu/KWh, having O&M costs of 5.3 mills/KWh, and
planned for 1,000 h of operation per year. Assuming a
3-yr construction and installation period and no future
inflation, such a turbine installation would require the
following charge components to cover the costs of power
generation:

Annual fixed costs	19.1 mills/KWh
O&M costs	5.3 mills/KWh
Fuel costs	30.3 mills/KWh
Total costs	54.7 mills/KWh

If a storage system of 75 percent efficiency could perform the same service functions at O&M cost rates of
1-5 mills/KWh and with an off-peak charging cost of 10
mills/KWh, then break-even capital investments could be
at a level that would require a 36-40 mills/KWh charge
for carrying the annual fixed costs. Typically, this
might allow a capital investment of $225-$250/KW for a
storage system of 25-yr life that could be built and installed in 3 yr, or about $170-$190/KW for a storage installation requiring 8-yr construction lead times. If
gas-turbine fuel costs were expected to increase to $4
per million Btu for the life of the storage system, the
allowable capital costs would rise to $340-$360/KW and
$260-$280/KW, respectively. Also, if the storage device
had technical characteristics that would result in one or
more of the utility system benefits discussed previously,
even higher capital costs of storage systems might be
warranted.

The capital cost of a storage system can often be bro-
ken down into a cost determined by the system's power
rating and a cost determined by the system's storage
capacity as follows:

$$K_t = K_p + K_s \cdot t_m$$

where

K_t = Total capital investment ($/KW)
K_p = Power-related investment ($/KW)
K_s = Energy related investment ($/KW)
t_m = Maximum discharge time at rated power (h).

For example, for conventional pumped-hydro installations, K_p includes the cost of the powerhouse with the turbines/ pumps and other associated equipment; K_s is primarily the cost of the reservoir.

In a previous example, a storage cost of $190/KW was shown to be warranted. Since K_p is about $140/KW for pumped hydro, the allowable costs for a reservoir with sufficient storage for a 10-h discharge at design power are about $5/KWh. This is achievable with large installations.

In the case of batteries, K_p (cost of future AC-DE converters) will probably be around $70/KW. Assuming a 10-yr battery life and a 3-yr construction time, the allowable cost for storage capacity is approximately $(190-70)/KW = $120/KW. For a 6-h discharge time, this translates to an allowable cost of $20 per KWh of storage capacity, which cannot be achieved for present lead-acid batteries and is a difficult target for advanced batteries. If competitive fuel costs increased to $4/million Btu, the battery capacity costs could increase to about $35/KWh.

Concepts such as batteries, flywheels, and superconducting magnetic energy storage devices with high costs of specific storage capacity (K_s) are at an economic disadvantage if the duty cycle requires extended periods of discharge (or charge) at design power. However, these concepts tend to have relatively low power-related costs, and their use may become economical if shorter periods of discharge (peaking service) and a large number of uniform duty cycles are compatible with the operation of the electric power system. Also, capital cost credits due to beneficial characteristics could impact such systems particularly favorably. On the other hand, methods such as pumped hydro (both conventional and underground), compressed air-gas turbine systems, and possibly thermal energy storage have higher power-related costs but rather low specific storage costs.[2] This makes them logical candidates for extended periods of charge (including weekends) and discharge.

The optimum utility mix may involve a combination of several storage methods with different operating and economic characteristics--just as a mix of different generating equipment is presently used to decrease the cost of electric power. Thus, the identification of optimum use (including the various possible benefits) for storage technologies and systems must be a key ingredient in efforts to establish technical and economic goals for utility storage systems and to design the most relevant R&D programs.

CONCLUSIONS

The electric-utility industry offers a large potential market for suitable advanced storage systems. Such systems, properly sized for either daily or weekly storage cycles, could be alternatives to primary generation units for either peaking or intermediate generation modes. Installations may be feasible at central generating sites or, in dispersed units, for locations on transmission and distribution networks or at the customer's site. Overall, if suitable storage technologies can be developed and demonstrated, as much as 10 percent of the installed primary generation equipment needed by the electric utilities in the 1985-90 time period might be displaced by storage systems.

Storage capacities of potential interest will range from 15 MW for dispersed substation installations to several thousand megawatts for large central station locations. Units will need storage durations ranging from 2 h to 2 days and will have to operate over charge/discharge time ratios from 0.2 to 2.4. Such wide ranges of operating parameters will provide great latitude for technological innovations.

Much less latitude may exist when economic considerations are coupled with operational parameters. Storage-device concepts are not likely to receive serious consideration unless the future revenue requirements to cover their costs are less than those for providing comparable service from primary generation equipment. To the extent that overall savings resulting from fuel cost reductions are accompanied by higher initial capital investments, it is likely that substantially more than a break-even cost advantage will be needed for the favorable consideration of storage systems because of utility capital acquisition problems.

The potential benefits of utility storage systems are great enough to warrant an R&D program that would clarify some of the operational uncertainties regarding their use. Such a program should include:

- Establishing quantitative models of the effects and benefits of storage-capacity levels on service reliabilities and optimum generation mixes and reserve margins. These should incorporate detailed demand projections for the post-1990 period reflecting probable changes in utility load characteristics.
- Developing the benefits of dispersed storage as a function of utility geographical, land use, and demand characteristics. This study should be on a regional rather than an average national basis.
- Establishing the possible interactions between storage, load control, and heavier interties between areas of differing load characteristics.
- Relating dispersed storage to dispersed generation and total energy systems.

In parallel with such studies, there is a need for R&D support of promising storage concepts in order to ensure the timely availability of utility storage options with a wide range of operating characteristics.

OPPORTUNITIES

The residential and commercial sector of the economy consumed about one-third of the energy used in the United
States (including allocated electric-utility losses) in
1972. This energy, almost evenly divided between the two
sector components, was overwhelmingly used as thermal
energy. A detailed study of 1968 consumption patterns
showed that more than 90 percent of residential usage was
for thermal purposes (with more than 70 percent for space
and water heating only), and more than 75 percent of the
commercial consumption was for thermal purposes. On a
combined basis, more than 85 percent of the energy use was
for thermal purposes, with about 65 percent for space and
water heating only.[3] These percentages have probably not
changed significantly since 1968.

Since World War II and until recently, oil and natural
gas have been preferred for space and water heating. For
the past few years, however, because of their declining
availability, as well as a tripling in the price of oil,
a substantial increase in the installation of electric
space-heating units has occurred. The heating plants installed in new home construction in 1969 were about 65
percent gas, 10 percent oil, and 25 percent electricity.
In 1973, the oil units remained at about 10 percent, but
natural gas heating declined to about 47 percent of the
market, while electricity rose to 43 percent. These
trends, coupled with the ban on new gas hookups in many
sections of the country, make it probable that electrical
heating units in new residences currently exceed new gas
installations and will continue to grow.

Although it is technologically feasible to store electricity (e.g., in batteries) on the customer's site,

30

relatively few benefits accrue to either the customer or
the utility. The principal benefits will arise from the
storage of thermal energy, which constitutes the end use
of most of the consumed energy. Such benefits may include
substitution of more abundant coal and nuclear fuels for
gas and oil, more efficient use of electrical energy and
capital investment, and increased use of solar energy.

Thermal storage for heating has been used extensively
in Europe, largely in the form of ceramic storage materials in individual room units that are electrically heated
during off-peak periods.[2] The economic incentives at preferred comfort levels and for predominately centrally
heated homes have not existed to stimulate such storage in
the United States. Hot water storage tanks can also be
considered a thermal storage system, although they have
generally been sized to match water-use requirements rather
than for the storage function.

In most cases involving space conditioning, thermal
storage does not reduce the demand for supplied energy.
Although in some cases thermal storage can use outside
temperature variations to reduce energy requirements, in
most cases storage may actually increase the need for
energy because of storage losses.

Since oil and gas are readily stored in tanks or pipelines, no incentive exists to use energy rather than simple
fuel storage when these fossil fuels are still available as
the primary source of heat. The impetus for storage will,
rather, arise because of efforts to increase reliance on
more plentiful fuels (coal and nuclear fission), to make
more effective use of electricity as a thermal energy
source, and to increase the use of solar energy to reduce
the need for electricity.

The social and national benefits of reducing energy consumption, particularly the consumption of scarce natural
gas and petroleum fuels, will not be achieved in the resi-
dential/commercial sector unless the individual consumer
is motivated to act. The possible motivations that could
lead to the widespread use of storage systems are similar
to those that might stimulate other conservation actions.
These include reducing the size of space-conditioning
equipment, adding to the comfort of the occupant, improving space-conditioning reliability, contributing to the
general convenience of the user, or saving the user money.

Installing storage devices in residential and commercial
buildings would not appear to offer size, comfort, reliability, and convenience advantages. At best, these would
be neutrally affected from the viewpoint of the user.

Therefore, the increased use of storage devices will probably be stimulated only by the consumer's perception of the
prospect of reduced costs. Without sufficiently reduced
costs, the use of storage devices will require legislation
or regulations that impose or provide incentives for their
use.

The basic economic problem of storage systems is the
increased capital costs of the space-conditioning installation. Unless offset by reduced energy costs, the user
will see no net benefits. The prospect for savings is
enhanced by cost/benefit analyses calculated for the life
of the building. However, most residential and, more importantly, most commercial builders do not generally consider life-cycle costs unless required to do so by market
competition. Homeowners have generally been less concerned
with long-term savings than with initial costs.

It may be appropriate that the government subsidize or
provide incentives for storage systems that are prospec-
tively most cost effective and energy conservative to dem-
onstrate their effectiveness, reduce their costs, and
promote their use. Utilities should be encouraged to work
closely with product and system manufacturers and the government to undertake thermal storage demonstrations to aid
in determining their advantages and effectiveness for
residential/commercial applications.

GENERAL CONSIDERATIONS

A wide range of considerations need to be evaluated in determining the possible benefits of residential/commercial
storage systems. Conditions do not exist that will make
energy storage beneficial across the board. The wide
variety of building types, of functional use, and the wide
range in efficiency of storage designs and types suggests
that energy storage will have to be evaluated separately
for each possible application to determine whether its incorporation is justified.

Storage may appear to be most beneficial in those buildings and for those functions in which energy is not being
used efficiently. Storage is not a substitute, however,
for energy-efficient designs, but it may permit recovery
and reuse of energy or provide other benefits where efficiency is low. Energy-efficient designs include such
characteristics as adequate insulation, placement of win-
dows to permit effective use of solar energy in the winter
while screening it out in the summer, and the transport of

energy from areas of excess to areas that are deficient. Where such designs have not been incorporated, storage may appear to be beneficial.

As designs improve and as the energy efficiency of existing buildings is increased by appropriate modifications, the benefits of storage in the residential/commercial sector will be reduced. The benefits derived from storage will be principally related to substitution of electricity generated from the abundant fuels, to load management to minimize peak loads imposed on the utilities, and to better matching of energy generation of all forms with the demand curves. Obviously, the consumer motivations for storage will still be based on perceptions of cost savings as discussed earlier.

Type of Storage System

Many different kinds of storage systems can be used for residential/commercial buildings. The storage system may be used only with the heating cycle or for a combined heating/cooling function. It may operate with either air or water transport of the stored energy. The storage medium may be rocks or ceramic materials, sensible-heat liquids, phase-change materials, or the building structure itself. The energy input may come from resistance heaters, heat pumps, solar collectors, or fossil fuels (generally on a supplementary basis).

Some European nations have widely used unitary (room-sized) heating storage devices containing ceramic storage materials charged by electrical resistance heaters during off-peak periods. Charging of the storage unit is controlled by the utility through carrier signals on the distribution lines. Space conditioning is obtained by circulating water to room radiators or by fans mixing storage-heated air with room air. While the high-temperature storage device works adequately with resistance heaters, it is not compatible with heat pumps operating at lower temperatures. However, for installations where stored heat from resistance heaters can be used at an equivalent heat-pump coefficient of performance (COP) of 1, unitary storage devices could supplement heat pumps during the coldest periods.

While the European type of modular, high-temperature storage system may find some application in the United States, larger-capacity systems operating with central space-conditioning units will be more common. Therefore, retrofit storage systems may for economic reasons need central system capacity. Furthermore, new electrical heating systems will probably increasingly employ heat pumps to achieve higher overall efficiency. Greater storage capacity will be required to optimize the consumer-utility interface. Greater storage weight and volume will be needed because of the smaller storage temperature difference available from the heat pump. Therefore, storage systems adopted for use in the United States will probably be both larger and more complex than those used in Europe. Multiple storage containers may separate heating and cooling functions, collect building waste heat, and serve as heat sinks to maximize the off-peak use of heat pumps. Capacities may even be made great enough to provide seasonal storage for balancing summer and winter energy requirements. These possibilities indicate that a wide variety of storage-system types, sizes, and costs need to be considered.

Retrofit or New Construction

Storage systems are applicable to either new or existing buildings. In most cities, little construction is currently taking place. Even with a large amount of new construction, a generation could easily pass before appreciable energy or scarce fuel conservation would result from more-efficient energy systems in new buildings. Greater short-term benefits could be realized from retrofiting existing structures with storage systems. Although the technical and economic problems associated with retrofit programs are greater than for systems designed for new construction, development emphasis placed on systems applicable to the retrofit market will have greater overall returns.

If developments show promise, means for removing or easing economic and institutional constraints will be needed to encourage wide application of the storage systems. A study funded by ERDA is examining the barriers to practical energy conservation retrofit opportunities. These findings may apply, or could be extended, to cover thermal storage opportunities as well.

Integral or Add-On

Storage systems may be an integral part of or separate from the basic heating/cooling system with which they function. Many of the same factors that apply to new construction and retrofit systems will also differentiate integral and add-on systems in terms of R&D emphasis.

Add-on storage devices are those that could be added to conventional thermal systems without major change. The basic system would be similar or use the same components regardless of whether or not storage is included. Add-on devices, by their nature, will be more adaptable to retrofit installations unless the devices have unsuitable volumes or require unsuitable locations. Integral systems, on the other hand, will be less adaptable unless the entire system, with storage, is sufficiently similar in volume, shape, or function to replace the original installation.

Diurnal or Seasonal

The primary characteristic of a seasonal storage system is the very large capacity that is required--on the order of a hundred times the daily storage capacity. The heat losses become very important for such long-term storage. More care must be taken to prevent thermal losses in a seasonal system than in one for daily storage.

While diurnal systems can generally be installed within the building, seasonal storage requires such large volumes that separate or additional locations may frequently be required.

The capital costs associated with the size and insulation necessary for seasonal storage systems have generally kept them from becoming economical. New technologies, increased energy costs, and the general desire to conserve scarce energy sources may warrant a review of seasonal storage systems and some development activity such as the ERDA-supported work on an "Annual Cycle Energy System." However, diurnal or short-term storage will probably find broader application and will have greater national impact at any given time.

Individual or Aggregate

A size relationship similar to that for diurnal/seasonal applications results from consideration of individual versus aggregate use of storage systems. When individual units are aggregated into a system large enough for several buildings, the total storage volume will be the sum of the individual units. With the larger volume, the lower surface-volume ratio of the aggregate unit reduces the thermal losses for indentical storage periods.

Lower unit-storage costs should occur as size is increased by the aggregation of individual systems. Depending on building density, however, thermal transmission costs may wipe out the savings arising from aggregation.

Control may be more difficult when a single aggregate storage system is used for several buildings. If individual accounts are maintained, cost billings would be difficult. Allocation of energy costs in an aggregated situation will only be feasible if a fair and inexpensive way of prorating consumption exists.

Preferences for individual or aggregate systems will depend on the circumstances of each particular case. However, in the case of single family residences, characterized by a uniqueness of ownership and a wide diversity in consumption patterns, aggregate systems are unlikely to be favored.

Solar Energy Storage

Because of the intermittent nature of solar energy and possible mismatches between supply and demand, thermal storage is an essential feature of solar-thermal systems. The energy from solar collectors tends to be at low temperatures and requires a large storage mass when stored as sensible heat. Although the efficiency of solar collectors increases (and the collector cost probably decreases) as the temperature of the collector output reduces towards the space-conditioning comfort range of 70°-75° F, the overall mass and volume of the storage device further increases. For almost all locations, however, a space-conditioning system using solar energy will probably have to be supplemented by an electric- or fuel-powered auxiliary energy source. Thus, optimal and synergistic use of both energy sources via thermal storage is important in designing building structures and thermal storage systems.

A major question is the quantity of solar energy to be stored. The storage system must be adequate to supply heat not only during the night, but for several consecutive cloudy days as well if complete independence from external energy sources is desired. In some regions, sunless winter periods are so long that complete independence is not feasible. Furthermore, the solar-collector system, to achieve independence, must be sufficiently large to heat the structure at the same time that it is storing more heat for the next sunless period. The sun may be overhead only one-third of the time during northern winters. Under such conditions, the capital costs of the combined collector and storage system may effectively limit the extent to which the solar system can economically supply the needed energy.

If an alternate heating system with another energy source must be included with additional capital costs as a backup for extended sunless periods, then solar energy use may be even further limited. However, if the storage system is economically justified, completely or in part, because it takes advantage of low off-peak electricity prices, then storage of solar energy, additionally, will enhance that benefit.

The energy from solar collectors need not be used directly in the space-conditioning system. A promising application is to use it as a source for a thermal heat pump in a solar-augmented heat-pump system. For such uses, the solar-collector outlet temperatures can be lower than with direct heating, thereby increasing the efficiency and probably reducing the cost. Additionally, the higher-available source temperatures for the heat pump will increase the COP and reduce the electricity consumed.

How the solar-augmented heat-pump system operates depends, in part, on the energy storage provisions. With minimum input storage, the solar collector only improves the heat-pump efficiency during sunlight hours. With greater storage, the solar input also provides a reservoir of higher source temperatures for heat-pump operations during sunless periods. If the overall system is designed to limit the heat-pump operation to off-peak hours, then a dual storage system is necessary--one system for the solar input storage and a second to store the heat-pump output energy for round-the-clock use in space conditioning.

Utility Impacts

The increased use of electrical space heating has not yet had an adverse effect on most electric utilities, because the latter have the capacity to supply the winter heating market--about 65 percent of the utilities have summer demand peaks.[1] However, electric heating can have a major effect on the peak-load growth of a winter-peaking utility. A summer-peaking utility only experiences a peak-demand increase of between $2\frac{1}{2}$ to 3 KW for each new residential air conditioning customer, whereas a winter-peaking utility adds a peak-demand load of from 10 to 12 KW for every new electric heating customer.

Compared to straight electrical resistance heating, the energy savings derived from heat pumps depend upon the severity of the climate. Current heat pumps use a simple cycle and are sized to operate efficiently in the temperature range where they operate most frequently. This design keeps the initial cost low and reduces the annual use of electricity, but results in a rapid drop in efficiency and output when the ambient temperature drops. Therefore, current heat pumps require supplemental electrical resistance heating under the coldest conditions (below about 30° F). This aggravates the supply problems of winter-peaking utilities. The supplemental heating adds to the load when the utility may already be experiencing a peak, yet use of the heat pump decreases the utility's annual revenue by reducing the load during milder weather when the utility has adequate generating capability.

Thermal storage, alone or in combination with other supplemental heat sources, including solar energy, could reduce the winter peak loads. Integrating storage, supplemental heat sources, and the basic residential heat pump in a system that would optimally use all components at an acceptable first cost will be important. If a satisfactory storage system can be developed, some of the capital otherwise needed for utility expansions might be displaced to the homeowner for a more effective and efficient total-energy system.

Thermal storage is essential for load-leveling. This is not restricted to electrical-resistance heating applications; thermal storage offers advantages for all energy sources.[4] This is especially true when the use of multiple energy sources (e.g., solar-supplemented electric space-heating systems) is considered. The resource savings and economic advantages resulting from the application of thermal storage to the full range of energy resources now used or expected to be used in the future for space conditioning can be significant.[5] If a diversity of thermal-storage systems were available, ranging from room level to central building systems, space-conditioning applications that could be operated without the consumption of scarce fossil fuels would be enhanced. This could be done with little or no disruption of life-styles and with a maximum of personal comfort.

Economic Factors

While some thermal-storage systems are currently available for residences and commercial buildings, their high capital costs deter their widespread use. Future storage systems, suitable for space-conditioning systems operating only with off-peak electrical energy, may well be more complex and expensive. Furthermore, the capacity and capital costs of heating and cooling equipment must be further increased if the system is to be limited to a duty cycle falling within the off-peak periods.

The capital costs must be compensated for by energy cost reductions before potential users will install storage systems. Reduced utility rates for off-peak consumption are not generally in use today. The price differential that might be available between on-peak and off-peak rates to offset the capital costs of storage systems has not been determined.

The differential will need to be substantial if the storage system is to pay for itself in the normal economic frame of reference of the user. An unsophisticated home-owner might expect to break even between direct loan re-payments and energy cost savings. At current interest rates, this would require energy savings ranging from $100 per year per $1,000 of storage-system cost for a new system in a new home with a 25-yr mortgage to about $200 per year per $1,000 of storage-system investment for a retrofit installation financed by a 7-yr home-improvement loan.

As new construction and remodeling of existing buildings reduce future energy requirements for space conditioning, the savings from only using off-peak energy will diminish. Similarly, the use of more efficient heating and cooling systems, such as improved heat pumps, will reduce the value of off-peak power. These trends will make it more diffi-cult to offset the capital costs of storage systems by in-put cost savings.

PERFORMANCE CRITERIA

No standard residential or commercial building units exist that can be cited to establish quantitative performance criteria for thermal-storage systems. A building energy ratio (BER), representing the number of thousands of Btu/ft^2/yr of energy input, is frequently used as a comparative measure of energy consumption. A study conducted by the Real Estate Board of New York and the City of New York in 1972 indicated that the BER's for commercial buildings in the city ranged widely from 72 to 403, with an average of about 210. BER's for residences also vary widely, depend-ing on the type of construction, the quality of insulation, and the severity of the climate. Therefore, performance criteria for energy storage must necessarily be more qualitative than quantitative at this time. Further, as both new and existing buildings are made more energy efficient, the performance criteria for storage will not remain constant. Some effort is made below to recognize the desirability of higher energy-efficiency designs along with the requirements of storage systems.

Technical Criteria

Independent technical criteria for storage systems are difficult to establish since they are closely related to and generally affected by the economics of the resultant systems. Nevertheless, certain technical criteria are desirable, depending of course on appropriate trade-offs or balancing with other criteria.

Design

The system should be designed to interface with both heating and cooling systems to allow storage from the widest variety of energy sources in various combinations. This should include the use of operating strategies that take advantage of the large daily variations in outside temperature during spring and fall for heating as well as the use of cool night air during the summer for daytime cooling.

The storage system should be compatible with creative building designs, systems, and materials and with opportunities to use or transfer energy generated in the building. These features and applications include:

- The use of thermal storage to capture waste heat for reuse (e.g., from solar sources, lighting, mechanical systems, and so on).
- Storage and use of excess heat generated in interior spaces of office buildings. For example, this heat may be used to heat the perimeter of the building in the winter or be integrated with the chilled water system during the summer in such a way that it may be discarded, thereby reducing the summer peak load.
- Thermal transport systems, including heat pipes, to facilitate the effective use of building energy by transporting the energy to where it is needed.

● Integration of solar energy collection systems into the envelope of the building structure and the mechanical systems, particularly on new buildings, as supplemental storage to be used for leveling out the demand curve. Solar energy development for buildings has generally not taken advantage of possible synergisms from considering the building as a total system. Studies to evaluate the feasibility of integrating solar panels with the building skin may lead to the development of dual-purpose skins.

● Window materials or designs to absorb heat when wanted and reject, or at least screen, heat when not wanted.

● Use of buildings as thermal sinks to reduce energy consumption and even out the demand peaks for both heating and cooling.

● New materials for thermal storage that are not limited to salts, waxes, water, and rocks. The use of the earth, insulation, and building mass should be considered.

● New concepts for creating large areas of controlled environments by total or partial coverings (e.g., pneumatic roofs, tents, shading devices, skylights, and the like).

The storage system should be adaptable to and compatible with presently used heating, ventilating, and air conditioning systems. It should be capable of being manufactured by traditional methods and labor skills and marketed through the usual distribution channels.

Capacity

The storage system should meet its peak energy and power requirements with a minimum of auxiliary devices. In some cases (e.g., solar sources), auxiliary space-heating devices will be required. Conversely, the storage system should take advantage of solar sources to reduce reliance on alternate fuels. However, in other applications, such as electric heat pumps, sufficient storage capacity should be provided so that the heat pump is a base-load device during a predictable period of the year.

Size

The storage system should be as compact as possible and preferably be installed within the structure. It should permit reasonably easy expansion, for example, by being modular.

Efficiency

The storage system should be as efficient as possible, and should be designed so that energy losses are recoverable to the maximum feasible extent. Low storage efficiency will increase energy use and, possibly, the use of scarce fuels. The acceptable level of efficiency will depend on balancing the objectives derived by storage in the specific application and the energy and scarce fuel costs incurred by application of storage.

Safety

The storage system should be fail-safe, must operate safely, and must not create new hazards.

Installation

The storage systems should be capable of being installed by current mechanical and plumbing trades without lengthy and complex retraining. The installation should be simple and, if possible, permit the use of tools to which the trades have been accustomed.

Environmental Criteria

The basic design, materials, and operational practices that are used should not impair public health or the natural ecology. Materials should not be used that are toxic or dangerous if released or that could adversely affect the environment during manufacture, distribution, installation, or operation of the storage system.

Economic Criteria

The most important criteria, and those that will be the most difficult to meet, are economic. As pointed out earlier, unless there is a distinct cost advantage or the imposition of legislative or regulatory requirements for storage, widespread use of storage systems may not occur, regardless of their conceptual desirability.

Life-Cycle Costs

Storage systems should be evaluated for their life-cycle costs, including the total first cost and annual operating and maintenance, financing, replacement, and any other costs that may affect other parts of the building system. The success of residential/commercial storage may depend on the public acceptance of life-cycle costing. The costs should be calculated for the anticipated life of the system, with annual increments. A range of inflation assumptions should be incorporated in the analyses.

Economic analyses should also reflect a range of future energy costs. The spectrum should range from a continuation of current costs at one end to the maximum predictable increases at the other. There should also be multiple assumptions about other uncertain cost variables.

All economic evaluations should also assess the consequences of widespread use of the storage-system concept on the general economic vitality of the nation and on any other indicators that will enable R&D decision making to take place in the broadest socioeconomic and political context.

Payback Costs

Marketplace behavior indicates that relatively short payback periods should be an economic goal. Although attractive life-cycle benefits may be demonstrated, a favorable marketplace response to storage application is not assured because builders, investors, and owners usually think in terms of only a very few years.

The residential market is governed, in large part, by developers who build for sale rather than for investment. They have little incentive to incorporate initially costly features that will save energy over a long term. The committee recommends a payback criterion of no more than 2 to 4 yr for the residential market. Even this may be too long in today's market. However, this kind of payback period might be sufficiently attractive to owner-buyers to persuade builders to include energy cost-saving features in new homes.

For the residential retrofit market, on the other hand, where the occupant is the owner and decision maker, longer payback periods (perhaps even up to 7 yr) may be acceptable, particularly if "pay-as-you-go" financing can be provided. Obviously this is not well enough established to permit it to serve as a real market guide.

The payback period for the normal commercial market (space for sale or lease) should be from 2 to 4 yr. ERDA should consider financing in assessing the extent to which its storage programs may be adopted in the normal market process. Additional incentives to owners in the non-residential market would be required to raise the payback criteria from 4 to 7 yr. Large institutional owners may accept longer payback periods. The federal government will accept the longest payback period, as evidenced by recent General Services Administration (GSA) life-cycle criteria of as long as 40 yr in some cases.

Institutional Factors

The evaluation of storage systems for residential or commercial applications should assess the impact of inducements that might foster their installation. Such inducements fall into two categories: those that seek voluntary installation of storage technologies through incentives and those that force the use of storage through regulatory mechanisms without regard to economic trade-offs.

The financial incentives that might establish the economic viability of storage systems should be investigated and quantitatively compared to the attributed value of national savings in energy consumption, specific fuels, and critical materials and in reducing environmental impacts. The incentives might include tax credits or rebates, capital or operating subsidies, and loan guarantees. Since the needs of owners and investors vary, a number of alternatives should be examined.

If national energy supplies dwindle because of unforeseen events, governmental regulations might be issued setting minimum standards for the performance of heating and cooling systems. Such standards would probably apply to total systems rather than to the storage component, but the kinds of technical performance parameters that might foster the installation of storage systems should be considered in evaluating the need for storage R&D when normal economic competitiveness is not clearly evident.

45

CONCLUSIONS

Electricity has become the most prevalent form of heating for new residential buildings. Because of natural gas shortages, it is probable that this trend will continue in the future. The presently favored electrical-resistance heating systems will most likely be displaced by electrically powered heat pumps, which are substantially more efficient on an annual basis. At low ambient temperatures, however, the efficiencies of current heat pumps fall to resistance heater levels and the resultant demand load on the utilities becomes as great as for resistance heating. Thus, the energy supply problems for winter-peaking utilities may actually become worse, since the overall annual revenues will be reduced with no reduction in the peak-demand load.

Thermal-storage systems, charged by heat pumps during off-peak hours, would be one way of alleviating the peak-demand problem. Alternatives might include solar-thermal augmentation systems, which also require thermal storage or the incorporation of oil-fired supplemental heaters, which could obviate the need for storage systems.

If solar energy is to supply all the energy for residential or commercial heating, thermal storage is an absolute requirement. Storage would not be required if solar energy is only used when available and alternate energy sources are used at other times. In this case, however, the alternate energy source will have to supply the full energy load at times, and solar energy could not be depended on to reduce the peak demand.

The preferred storage system will largely depend on the comparative overall costs to the consumer of capitalization and operating charges of various alternatives. These will be a function of the as yet largely undetermined capital costs for storage systems as compared to supplemental systems and on the spread between on-peak and off-peak electrical energy charges.

A well-designed research, development, and demonstration program for thermal energy storage systems is warranted to establish such cost and break-even parameters. Such a program should recognize that the best and cheapest energy storage system is a well-insulated building! Storage will show the greatest cost benefits or scarce fuel conservation potential in those residential and commercial buildings where large amounts of energy are used, so that, in effect, energy storage may look best in energy-inefficient situations. However, as energy conservation promotes greater attention to more efficient design and reduced energy use, the importance and benefits of energy storage will decrease.

In general, new construction emphasizing energy-efficient designs should be more readily accepted by the institutional structure (i.e., costs can be included in the mortgage) and may cost less than supplemental energy storage systems. Therefore, ERDA should emphasize a spectrum of approaches to reducing energy requirements in residential and commercial structures as part of the thermal-storage R&D program so that thermal storage can be evaluated in the context of energy-prudent buildings.

However, new construction will have a relatively slow and minor impact on overall residential/commercial energy needs for many years. Therefore, a major R&D emphasis should be on thermal-storage systems that are compatible with and beneficial in existing structures. Such studies should also be conducted in the context of other cost-effective conservation retrofit practices that conserve energy.

In addition to conducting technical and economic evaluations of storage devices and systems *per se*, study should be concurrently carried out of the kinds and levels of incentives or other actions that might foster the use of thermal-storage systems to achieve national energy goals if consumer cost savings do not seem to be attainable.

4 INDUSTRIAL APPLICATIONS

OPPORTUNITIES

Industry consumes almost 40 percent of the total energy used in the United States.[3] Some industries, as a result of the continuous nature of their work processes, consume energy at a more stable, uniform rate than do residential or commercial customers. On the other hand, some major industries still use batch processes, with the demand for energy varying markedly during the process stages. Such variations in consumption patterns, together with the wide ranges of energy sources, operating temperatures, and cycle times, lead to a wide variation in storage opportunities.

Industry consumes energy for many different purposes. A 1968 study of industrial consumption patterns showed the estimates in Table 3.[3]

TABLE 3 Industrial Energy Use

Use Category	National Energy Consumption (%)	Industrial Energy Consumption (%)
Process steam	14.6	39.2
Space heating	2.1	5.6
Direct heat	11.5	30.8
Electric drive	7.9	21.2
Electrolytic processes	1.2	3.2
TOTAL	37.3	100

The complexity of determining storage opportunities is inferred by the above data. The process-steam and space-heating categories use energy for the most part at relatively low temperatures (less than 400° F). The direct heat and, in part at least, the electrolytic processes use high-temperature energy. The potential for storage at each temperature level will be different. As a result of these differences, as well as differences from industry to industry, a definitive analysis of storage opportunities for specific industries will require the participation of personnel intimately familiar with their characteristics.

Although energy requirements have always been an important aspect of industrial operations, until recently energy-related issues have been considered strictly on an economic basis. Energy costs were typically a small percentage of the product price (11 percent average in 1970). Currently, the high degree of uncertainty about future fuel availabilities coupled with higher fuel costs provide additional incentives for improving energy-management practices. The incentives include increased security of energy supplies, improved efficiency, and reduced costs. The greatest returns will probably come from the adoption of conservation practices that are now profitable, although storage systems have a potential role in responding to these incentives.

Security of energy supplies could be increased by switching demands from periods of uncertain supply to times when supply is more assured, or from a scarce energy source to a more plentiful one. Energy efficiency could be improved through the recovery and reuse, before rejection to the environment, of greater fractions of the industrial waste energy, thereby directly reducing energy costs.

Energy storage is frequently thought of as an add-on device in which recoverable energy can be accumulated for later use. In many cases, however, inherent storage systems already exist in the production process (furnaces, molten materials, chemical intermediates) that can be used by a rescheduling of process steps. The range of storage opportunities in industrial operations is typified by the following examples.

Seasonal Storage of Intermediate Product

In the chemical industry, process steam is used to evaporate inorganic chemicals from a low concentration to dryness. The steam is also used as a source of space heat in the winter. In one case, described to the committee by

the Dow Chemical Company (Midland, Michigan), this required
the yearlong operation of boilers (each having a back-
pressure turbine-generating electric power) to supply both
process and space steam. The electric demand was such that
the steam was generated as a by-product during the summer,
even though space heating was not required during that
period. The excess summer steam was vented.

The work process was modified so that the inorganic
feed solution was stored during the winter in underground
salt formations from which the brine had been extracted.
The evaporators were then primarily operated during the
summer months, using the previously vented excess process
steam for drying. As a result, the previously wasted
excess off-peak summer steam was "stored" in the product,
and the need for process steam was correspondingly reduced
during the winter months.

This example also illustrates another potential application for energy storage. The production of process
steam as a by-product of electric power generation, using
extraction or back-pressure turbines in a topping cycle,
is inherently an efficient process. However, the steam
and the electric power are not always needed in a matching
time sequence. A storage device might optimize the total
energy use.

Short-Term Storage
in Process Equipment or Storage Devices

An industrial plant having significant inherent storage
capability (e.g., furnaces, space conditioning, compressed
air systems) or to which short-term storage devices (batteries, flywheels, compressed air tanks) can be added can
operate with interruptions of energy supply for a few
minutes on a rotating or priority equipment shutdown basis
as needed to reduce load peaks. The use of such storage
concepts does not reduce overall energy use, but may be
very cost-effective in reducing electrical demand charges.

Low-Temperature Storage and Reuse of Waste Heat

Many processes generate intermittent energy at a wide
variety of temperatures. These may range from furnace
slag and molten metals at temperatures in excess of 2,000°
F to chemical products at less than 500° F. The energy
from the process streams must be removed to permit further

processing of products. This has typically involved re-
jecting the heat directly into the environment. Regenerative heat recovery was introduced only if convenient.

Storage devices would potentially permit the recovery
of the heat for use elsewhere in the plant. Thermal-
storage devices could use hot pressurized water (as in
Europe), low vapor-pressure refined oils, reversible chemical reactions, or heats of fusion. Potential applications
include the ceramics industries, where high-temperature
batch or semicontinuous drying and firing operations are
carried out, and the metallurgical industries, which also
use batch operations.

Industrial On-Site Electric Energy Storage

The on-site storage of electricity could increase the reliability of supply in case of short-term power outages
or, at the very least, supply emergency power to critical
operations such as process control computers or safety
systems. (This may become important if service reliability
is reduced due to reduced utility reserve capacity.)

On-site storage would also reduce the capital costs of
utility transmission lines and switch gear by providing an
internal supply of energy for peak loads. When demand
charges are high, an energy cost savings is also indicated.

For the economics of on-site storage to be attractive,
the load factor for the storage device must be high enough
to warrant its capitalization. Thus, the device must be
used regularly and be justifiable as a load-leveling or
other tool and only incidentally as an emergency power
supply. Economic justification solely as an emergency
power supply will be more difficult.

Interruptible Process Loads as Energy Storage

A number of processors (particularly in the electrochemi-
cal industries) purchase electricity on an interruptible
schedule at reduced rates. Such a practice is used by
utilities as an emergency precaution in case unexpected
outages of generation capacity or extreme peak loads are
experienced. However, extending the practice to other
industries with production flexibility may expand the use
of low-cost energy produced from readily available fuels
as compared to higher-cost fuels that are typically used
during peaking periods. In such cases, the product itself

would become the storage device. The added capital cost of product storage (plus equipment protection during enforced outages) would need to be justified by lower energy charges.

Conversion of Waste Products
into Storable and Usable Forms

In some cases, the storage of waste energy would be hindered by the form in which the energy is released by the process. For such cases, a chemical or physical conversion to a more readily storable form would enhance the opportunity for and the value of the storage process. In the case of gases released from blast furnace operations, for example, a change in the chemical form of the gases to raise the heating value from 100 to 1,000 Btu/ft^3 would ease the storage problem and increase the value of the stored energy. Similar examples can be found in the paper industry, where by-products can be converted into more readily storable and consumable forms.

PERFORMANCE CRITERIA

The range of opportunities for industrial storage applications is too great to delineate specific performance criteria that will predict the commercial viability of R&D proposals. Some of the more promising opportunities will come from the exploitation of "inherent" storage capabilities associated with specific equipment or processes. These cannot be generalized, and storage concept proposals for such purposes will need the critique of knowledgeable experts from the potential user industries to ascertain their worth.

Add-on storage devices will, by their nature, have a more general base of applications. Even so, the range of performance parameters of potential interest will be very broad. Typical performance parameters can, however, be described.

Technical Criteria

Capacity

The unit of capacity will depend on the type of energy being stored. The general types that will be considered

are thermal and electric. Thermal storage can be further
subdivided into the storage of high-grade excess process
steam and the storage of low-grade waste energy.

The capacity requirements for storage of process steam
will depend on the excess steam-flow rates, which might
vary from 100,000 to 1 million lb per hour, and the period
of the process swings, which might be assumed to average
about 5 min. For an average of two swings per hour, this
would give an average of 25 min of charging to the storage
system. Using these averages and the assumed limits for
excess steam flow, the storage capacities would range from
about 10 million to 100 million Btu (3 MWh to 30 MWh).*

The storage of waste heat leads to even more tenuous
capacity estimates, since they will depend upon the magnitude and temperatures of waste heat sources. From a practical viewpoint, since primary heat is currently available
at an upper cost limit of about $2.50 per million Btu,
industry will have little incentive to invest in storage
equipment for small quantities of waste heat.

Storage systems for waste heat, therefore, should have
storage capacities of no less than about 100 million Btu
(30 MWh). Maximum capacities might range to 10 times that
size. Further, whereas high-grade process steam heat may
be stored and used in the same local area, the low-grade
heat content of waste heat frequently makes it unusable in
the process areas from which it is collected. Consequent-
ly, transportation in space, as well as storage in time,
may be essential to give waste heat any economic value.

The capacity requirements for electricity storage will
depend upon the size and nature of the industrial plant as
well as the function that the storage is to serve. The
capacity power levels will probably fall in the range of
1 to 10 MW. The energy capacity may range from a few
kilowatt-hours for load-leveling functions in small diversified plants to perhaps 100 MWh for major load-leveling
and emergency-power functions.

Charge/Discharge

Most thermal-storage applications will be characterized by
relatively long charging times as compared to discharge

---

*The parenthetical numbers are metric equivalents of the
energy units generally used in industry. See Appendix B
on units and conversion factors.

times. As a consequence, the energy charge/discharge rates will probably fall in the range of 1:5 to 1:4. An exception may occur in storage systems for waste heat where, depending on the use made of the stored heat, the charge/discharge rates may approach 1:1.

Storage of electricity will generally be accompanied by relatively fast, high rates of discharge at infrequent intervals. The energy charge/discharge rates may, therefore, range to higher values than will thermal storage systems--up to 1:20 to 1:30.

System Life

All industrial storage systems will need to last a minimum of 10 yr, and even longer lifetimes would be preferred. The devices should be capable, however, of adapting to installation alterations to accommodate the changing products and processes of dynamic industries.

Size

Industrial storage systems will probably have no limiting dimensional or weight parameters.

Safety

Industrial storage will be intimately associated with the industrial processes that, in some cases, involve the handling of hazardous materials. Evaluation of the added probability and consequences of accidents that might be caused by having the energy storage on-site will be necessary.

Environmental Standards

Industrial storage devices will be subject to the same environmental constraints as the industries in which they are installed. Thus, special standards are not foreseen.

Economic Criteria

The economic trade-offs between operating savings and capital outlays should be made on a discounted cash flow basis, with the discount factor appropriate to the industry concerned. By virtue of obsolescence and other risk factors in industry, the discount factor may be considerably higher than for electric utilities.

Care should be taken that a realistic market be identified for the stored energy. As already mentioned, this might exist within the plant or might involve transportation of the stored energy off-site. The costs of transporting the stored energy to the point of use, including losses during transportation, should be included in the trade-off analysis.

SPECIAL CONSIDERATIONS

Because of the diverse characteristics of potential industrial storage applications, potential users must be identified and their assessment of proposed storage techniques obtained prior to the pursuit of costly R&D projects. As a general rule, storage projects in the R&D stage should not be planned solely for a single application, but should be designed for general use. This will require a greater versatility in design flexibility than will be the case for electric-utility storage concepts, for example. Such flexibility will also be essential to enable the storage system to be adapted to changing products and processes over the useful life of the installation.

The larger industrial organizations may have the knowledge, motivation, and financial capability to develop and apply "inherent" storage concepts without the need of federal R&D assistance. On the other hand, many smaller companies and industries will probably not have the required levels of awareness, expertise, or financing to implement or optimize inherent storage concepts, even if they promise longer-term operational and economic benefits. Therefore, a definite role for ERDA exists in identifying and disseminating the basic opportunities and probable benefits for "inherent" storage concepts and approaches.

With regard to add-on energy storage devices, even the larger companies may not have as yet assessed them, nor even be fully aware of current options and relevant developments in storage technology. ERDA's role, therefore, should include:

● The identification of significant applications and benefits of add-on industrial storage systems through a dialogue with industry.

• The conduct of specific analytical studies to quantify and verify benefits.

• The support of selected demonstration programs to prove practical applicability and to reduce the technological risk to industry.

CONCLUSIONS

The industrial sector is the major energy user in the United States. Much of the energy is discarded as waste heat, either because of the intermittent nature of production processes or because temperatures have dropped too low to be of further use on-site. Current energy costs are at a level, however, that should provide economic incentives for the adoption of improved energy-management practices that could result in significant energy savings. Storage systems have a potential role.

Industry is so diverse and complex in the ways in which energy is used that the exact form and role that storage might take is not clear. In many cases, the capability exists for "inherent" storage concepts in which the equipment, the processes, or the product itself may be useful as a short- or long-term energy accumulator. In other cases, the employment of add-on storage devices, particularly for low-grade waste heat, would be indicated if the stored energy can be economically transported and utilized on- or off-site.

This complexity militates against the desirability of a centralized R&D program that would be aimed at the development and demonstration of specific storage devices for use by industry. Rather, ERDA should:

• Establish a mechanism for conducting continuing dialogues with a wide range of industry representatives to determine significant storage opportunities for both inherent and add-on systems.

• Conduct specific analytical studies to quantify and verify potential benefits.

• Arrange for wide dissemination to industry of information on storage applications and technologies.

• Support selected demonstration programs to prove the practical applicability and to reduce the technological risk for industrial storage systems.

5 TRANSPORTATION APPLICATIONS

OPPORTUNITIES

Most transportation systems in the United States at the present time depend upon on-board storage of the total energy needed for propulsion and the operation of auxiliary devices. The principal exceptions are pipelines, conveyor belts, and electrically operated trains and trolley cars drawing power from the rail or from an overhead conductor. These exceptions present no significant energy storage opportunities except for emergency power outage or special requirements to deviate temporarily from the power rail or conductor. Flywheel energy storage has been proposed for this latter application.

The remaining transportation systems now depend upon petroleum-based liquid fuels. Indeed, some of the older transportation modes, such as railroads and ships, have essentially completely converted from coal to petroleum in the last 40 yr.

The development of alternatives to petroleum for providing transportation energy could diminish the demand for liquid fuels. About 30 percent of the national energy consumption in 1972 was for transportation purposes.[6] At the same time, the dominant role of petroleum in transportation is an indication of its outstanding suitability--its convenience, low mass and volume per unit of delivered energy, and overall economy.

Although many alternatives have been examined in the past few years, none can match all of the favorable characteristics of petroleum-based fuels. Some well-known examples are:

• Battery-stored electricity cannot presently power vehicles to the speeds and ranges of gasoline vehicles

56

because of the large masses and volumes needed by the storage batteries. Nevertheless, the convenience, low pollution, and low noise of electric vehicles already make them attractive for some special uses, such as short trips with many stops, in which gas vehicles would usually be left inefficiently idling. The attractiveness of electric vehicles may increase if the costs of petroleum fuels increase and as lightweight batteries are developed.

● At the other extreme, nuclear fuel offers a maximum potential for energy per unit mass and has therefore been studied frequently for aircraft and marine applications requiring extremely long ranges. The inconvenience, crash hazards, and lack of public confidence and acceptance of nuclear fuels have been the major barriers to its commercial adoption, although its technical feasibility is accepted.

● Intermediate transportation possibilities might be achieved with chemical storage fuels, such as hydrogen or methanol, which have been proposed for use in aircraft and road vehicles. Both hydrogen and methanol offer some advantages over gasoline, but at the present time these advantages do not offset the economic and convenience factors inherent in the widespread distribution system that has been developed for gasoline. If petroleum becomes scarce and more expensive, some experts and observers envisage a changeover to a "hydrogen economy," in which the uses of hydrogen for storage and transmission of energy will combine with its many industrial applications to provide a potential alternative to petroleum for transportation systems.

Many conceptual designs use these and other alternative sources of energy. Additionally, hybrid vehicles can be conceived in which more than one storage system is used to combine the inherent advantages of different devices at different points in the use cycle. Some examples of such hybrids include:

● Recent experiments at the Jet Propulsion Laboratory show that a combination of hydrogen and gasoline is less polluting than straight gasoline as an automobile fuel.

● A hybrid vehicle for urban use that combines a flywheel with a small gasoline engine can be operated with significant fuel savings compared to the conventional car. Depending on the design, the gasoline engine might be operated only intermittently to charge the flywheel or continuously with the flywheel supplying part of the starting and acceleration energy.

• A hybrid vehicle that combines electric drive from a storage battery with a small gasoline engine that operates efficiently and steadily at cruise power to continuously charge the battery has been proposed as a means of reducing gas consumption while retaining the ease of recharging associated with refilling the gasoline tank.

• The combination in an electric vehicle of battery storage with a flywheel to handle the peak power demands and to absorb the deceleration energy may have some performance advantages over a simpler vehicle powered only by a battery.

The establishment of detailed performance criteria for all possible transportation energy storage devices is clearly a formidable task. Any attempt to generalize requirements carries with it the possibility that appropriate criteria for one set of storage devices for one application might misstate the requirements for another. Thus the analyses have focused on only two types of storage devices: the storage battery for electric vehicles and the flywheel for hybrid gasoline-engine/flywheel vehicles. These appear to be the principal vehicle alternatives capable of challenging the conventional automobile for a sizeable niche of the transportation market in the United States.* The development of criteria for these classes of storage systems will indicate the factors that will be important in other devices if they are to achieve market acceptability.

Electric Vehicles

Electrically powered vehicles have been in limited but growing use for some years in England, Germany, and Japan, as well as the United States. At present, the lead-acid battery is used universally as the storage medium. The low-energy densities restrict the vehicles to uses requiring very short ranges--golf carts, urban delivery vehicles, and a few municipal buses. Intensive research activity in a number of government and private laboratories is aimed at the development of batteries with higher specific energies (sodium-sulfur and lithium-sulfur are

---

*Other nations, with different transportation system mixes and problems, may benefit from different vehicle storage systems. Some are placing greater R&D emphasis on all-flywheel systems, for example.

examples) that may enable electric vehicles to attain higher speeds and greater ranges.

Because markets for electric vehicles already exist, the performance criteria for storage batteries that will lead to wider acceptance of electric vehicles must be regarded as time-dependent. The early criteria, applicable to increased uses of the present generation of vehicles, will reflect the requirements of a very narrow segment of the potential market. The performance of new storage batteries should stimulate the expansion into that market.

The special requirements for battery recharge (discussed below) and the lack of readily available service facilities equivalent to the familiar neighborhood garage imply that for the next decade the electric vehicle will only be used in fleets operating on short routes and serviced at a common center owned by the fleet operator. An example is the postal van fleet operated by the U.S. Postal Service (USPS).

The USPS currently has 135,000 routes, of which about 30,000 are suitable for electric van service, even with the restricted performance offered by lead-acid batteries. A first experimental fleet of 350 electric vans has just been delivered and is in service at two localities in the country. Although the first cost of the vans is almost twice that of the corresponding gas version, the fuel and maintenance costs have proven to be so low ($123 per year per vehicle for electric power and $139 for maintenance) for the initial operations that the projected life-cycle cost is competitive. This small fleet operation could conceivably be followed by other fleet operations by the government and by major commercial operators, such as the telephone companies, with similar short-range, defined-route requirements.

If the early applications live up to expectations, taxicab fleets might be the next candidates for electric vehicles. An increase in battery performance will probably be required first, however. Since the present 118,000 cabs in the United States average 250 miles per day each, several recharges per day would be needed with present lead-acid batteries. Unless the waiting period at taxicab stands is used for recharging, an electric taxicab fleet must await the next generation of storage batteries to achieve the needed range. Additionally, techniques will have to be developed to enable the taxicabs to provide heat, air conditioning, and other service comforts presently available to the riders.

Battery-powered municipal bus fleets will encounter the same problems of range and comfort as the taxicabs. A current Japanese program is experimenting with the rapid change of the entire battery unit as a method for operating a bus fleet with lead-acid batteries.

Fleet operations of the type discussed above represent only a small fraction of the vehicle market, both in numbers of vehicles and in fuel consumption. Electric vehicles will need to penetrate the major market represented by the privately owned automobile if they are to have a sizeable impact on future national energy requirements. The private automobile accounts for about 80 percent of the fuel consumed by all domestic passenger transportation modes and for over 90 percent of all passenger miles traveled.[6] The dominant factors determining the penetration will be costs, performance, and flexibility (method of recharge).

The Federal Power Commission (FPC) in 1967 estimated a maximum electric vehicle penetration of the private car market at 2-4 percent or, roughly, 200,000-400,000 vehicles per year.[7] This survey predated the oil embargo, which increased concerns about fuel availability and costs. An updated evaluation is needed to measure the impact on the potential market of these increased concerns as well as the growing concerns about environmental pollution. It is significant, however, that the survey assumed the same first cost for electric and gas vehicles. Since first cost is still apt to dominate individual purchase decisions, competitive vehicle costs will be essential.

Battery performance limitations during the next decade will restrict the electric vehicle to shopper/commuter ranges of about 50 miles between recharges. New battery capabilities will have to be developed to raise the electric vehicle range and capacity to those commonly associated with family car performance levels--4-6 passengers with 200 miles between recharges.

The recharge procedures for the individual owner will be very different from the familiar "fill-er-up" at the local gas station and may constitute the greatest obstacle to rapid adoption of the electric vehicle, although acceptable costs and performances will also be difficult to achieve. To get "gas station" recharge convenience, even with a 15-min recharge period, would require 1,000-2,000 ampere currents at 115 volts. This would impose impractical technical requirements on both the vehicle battery and charging station, so much so that this rapid recharge rate is not considered a reasonable performance criterion.

An alternative that would give the driver somewhat the
same convenience and driving flexibility would be that of
quickly replacing the entire discharged battery pack with
a freshly charged one. The techniques for doing this
rapidly could probably be developed. A whole set of institutional problems would have to be solved, however,
including financing and ownership rights for the exchange
packs. New solutions might be forthcoming if widespread
use of the electric vehicle seems assured, but in the absence of such assurances, no basis exists for postulating
this technique as a positive market force.

For the foreseeable future, the preferred method of recharge appears most likely to be a relatively slow overnight process at a charging rate of 5-20 watts/lb of battery weight. If excessive discharges require charging at
other times, they must be treated as emergency breakdowns
of the vehicle. A reliable state-of-the-charge meter for
the vehicle will be imperative for this approach.

The inclusion of regenerative braking in the electric
vehicle would extend the effective life of the battery
between charges by converting vehicle kinetic energy into
charging electricity during decelerations, but special
control circuits would have to be incorporated to prevent
excessive current rates. During normal stops, the regeneration in the speed interval from top speed to 15-20 mph
would permit recovery of most of the kinetic energy. During panic stops, the hydraulic brakes would take over and
a current limiter would dissipate the kinetic energy and
none would be recovered. Such a system has been installed
on the electric postal vans at modest cost.

Even if electric vehicles (and battery storage systems)
can be developed that will have a wide market appeal, the
impact on the total transportation consumption of energy
will not be marked. Figure 3 presents vehicle efficiency
estimates for several fuel scenarios, including all energy
losses from the point of extraction of the fuel from the
earth. The left-hand boxes compare component and overall
efficiencies for the consumption of petroleum, whether
used directly to fuel internal combustion engines or to
generate electricity that is then employed to charge an
electric vehicle battery.

For the assumed internal combustion engine/drivetrain
efficiencies, which are about the same as the most efficient current automobiles and somewhat less than are expected to be achieved as a result of legislated future
mileage standards, the electric vehicle is slightly less
efficient than the gasoline-powered vehicle for the

selected driving conditions. The federal driving cycle
is designed to evaluate automobile emissions and does not
necessarily correspond to a typical driving cycle.

The right-hand boxes may be more representative of fuel
availabilities by the time an electric family car becomes
widely available. In these cases, coal is assumed as a
basic fuel source, and the comparison is between vehicles
powered by gasoline refined from synthetic coal-derived
liquids and vehicles powered by electricity generated from
coal. Because of the anticipated inefficiency of coal-
liquefaction processes, the overall electric vehicle system
will be about 50 percent more efficient than will be gasoline vehicles.

System efficiencies, such as those shown in Figure 3,
do not afford a true measure of energy consumption, however.

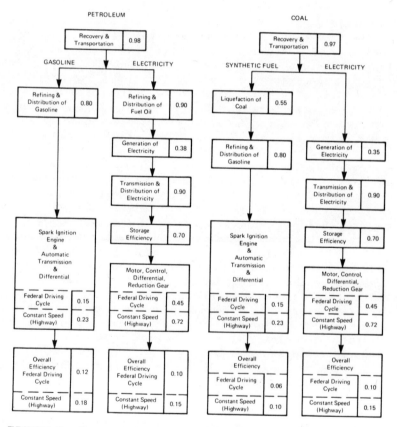

FIGURE 3 Comparative transportation efficiencies.

Detailed analyses of vehicle configurations have shown
that even with advanced batteries the mass of electric
vehicles for the same payloads will have to be greater
than their gasoline-powered counterparts. This mass
difference increases rapidly as the required range is
increased. When the effects of the greater masses are
evaluated, the total energy consumptions for the electric
vehicles will increase relative to the gasoline vehicles.[8]
As a consequence, the apparent superiority of the electric
vehicle for coal-based fuels will largely disappear, and
the gasoline vehicle will become substantially more fuel-
effective when the fuel source is petroleum.

Although overall energy savings may not materialize from
the use of electric vehicles, their development may decrease
the need for massive future investments in synthetic fuel
plants by making coal or nuclear energy, in the form of
electricity, viable transportation fuels. In addition,
the adoption of electric vehicles on a broad scale would
materially reduce urban pollution problems.

Hybrid Flywheel Vehicles

Electric vehicles face formidable technical and financial
problems that make achieving the performance characteristics required to secure broad market acceptance very
difficult. In particular, the vehicle will probably be
immobilized for several hours to recharge batteries after
relatively short driving ranges, thus severely limiting
the operational flexibility as compared to a gasoline-
powered vehicle. Consumers desire the essentially unlim-
ited range of gasoline-powered vehicles, supported by a
universal supply and service infrastructure. This desire
may well postpone the emergence of the electric vehicle
for other than special purpose uses unless and until gasoline becomes restricted in availability.

The gasoline vehicle is not without its problems, however.
The aggregated fuel requirements are enormous and are aggra-
vated by the mileage losses associated with pollution-con-
trol devices needed to meet legislated emission standards.
In part, both the low average miles per gallon and the pol-
lutant emissions are a result of the present need to install
engines sufficiently powerful to meet the maximum demands
during starts, accelerations, and high-speed cruising, even
though the bulk of the average driving cycle requires only
a fraction of the rated horsepower. The engines, therefore,
do not operate at their maximum efficiency and minimum
emission conditions most of the time.

TABLE 4 Performance Criteria for Electric Vehicles

	Delivery Van	Shopper/Commuter	Bus	Family	Hybrid Family Battery + Flywheel
Vehicle					
All up weight, lb	3,500	2,000	10,000	4,000	4,000
Max. speed, mph	35	50	40	55	55
Max. range, miles	30	50	100	200	200
Stored energy, KWh	25	30	150	150–200	150–200
Batteries					
Weight, lb	1,800	600	3,000	1,000	800 + 200
Volume, ft^3	5	4	30	6	6
Wh/lb	14	50	50	150–200	190–250
Max. discharge, watts/lb	14–28	83	40	120	40, 900
Max. charge, watts/lb (overnight charge)	2	5	40	20	20
Life, yr	4	5	5	10	10
Life, 1,000 miles	30	30	250	100	100

A flywheel energy storage system combined with a smaller engine has been proposed as a promising solution to these problems, while retaining the desired range and flexibility characteristics of the gasoline engine.[9] Two different concepts are receiving attention; both would combine the flywheel and a small gasoline engine in a hybrid system.

In one concept, the flywheel would be periodically recharged by the gasoline engine (which would be turned off between charges). The motive power for the vehicle would be drawn from the stored energy from the flywheel. The second concept would use a small gasoline engine as the primary motive power while drawing supplemental energy from the flywheel for high power needs. The flywheel would be kept charged through a power takeoff from the engine. Both concepts would also use the flywheels for regenerative braking, thereby reducing the energy input needed from the engine.

Significant improvements in urban fuel economy are projected for hybrid power systems. As much as 100 percent improvement is possible, with little loss in vehicle performance. Further, the operating range of the vehicle would only be restricted by the size of the fuel tank. Because the gasoline engine would be sized to operate at or near its best design point, pollution control will also be aided.

The full realization of these concepts in a practical, marketable vehicle has not yet been demonstrated. Further technical developments are needed in such component areas as the flywheel itself (materials, fabrication methods, suspension system, vacuum housing, and seals) and the continuously variable transmission connecting the engine, the flywheel, and the drivetrain (efficiency and noise). Manufacture of the flywheel systems at competitive cost levels has not yet been demonstrated. Importantly, proof is still lacking that the flywheel installation will be safe.

When and if these problems are satisfactorily resolved, the hybrid engine/flywheel system may have a promising future as a propulsion system for advanced generations of vehicles.

PERFORMANCE CRITERIA

Electric Vehicles

Table 4 summarizes the technical requirements batteries would need to meet to deliver assumed performance levels

for a range of vehicles. The characteristics for delivery
vans are attainable with current lead-acid batteries. The
shopper/commuter vehicles and municipal buses might be
feasible in about 10 yr, based on current programs for advanced batteries. Because of the major problems in lightweight battery developments and the need for individual
owners to become accustomed to driving limitations imposed
by recharging, the high-performance, gasoline-powered
family car is unlikely to be replaced by an electric car
in less than 25 yr. Table 4 also includes, as an example
of a hybrid vehicle, an electric family car in which the
battery-powered drive is supplemented by a small battery-
charged flywheel to provide for the infrequent high-power
excursions in the driving cycle. The flywheel performance
characteristics given in the last column would be sufficient to provide full power for a period up to 1 min. The
flywheel markedly decreases the maximum discharge rates
required from the battery, but it also increases the required energy density in the battery for a fixed overall
weight.

Size

The size and weight of the storage battery must be compatible with the vehicle. Total volumes of 4-6 ft^3 for
personal vehicles and up to 30 ft^3 for buses are realistic.
The weights are critical to performance. For low-speed,
low-range delivery vans, up to 50 percent of the vehicle
gross weight might be assigned to the batteries. As performance expectations increase, the fraction of battery
weight must decrease. The highway speed, high-range family
car cannot allocate more than 25 percent of the gross vehicle weight to the battery sets.

Storage Capacities

To achieve the requisite weight limits, the specific energy capacity of batteries (Wh/lb) must increase as the
range increases. Present lead-acid batteries have an
energy density of about 14 Wh/lb. The shopper/commuter
vehicle will need densities of about 50 Wh/lb, and the
family vehicle will require about 150-250 Wh/lb. Power
densities will also have to increase over current lead-
acid values. More than 100 watts/lb will be necessary

for family vehicles unless hybrid systems are incorporated
to reduce peak power rates.

Charge/Discharge

The discharge rates required of batteries will depend on
the driving cycle. The charging rates will, on the other
hand, be established by the recharging mode. Assuming
overnight charging of 7-10 h to use off-peak utility
capacity, the charging rates will be much less than the
maximum discharge rates, except for buses. The maximum
charging rate for the family vehicle for a full charge
over a 10-h period would be about 20 watts/lb. A 1,000-lb
battery set would require a charging device of about 100
ampere capacity at 220 volts.

Lifetime

The described battery characteristics assume that the
energy discharge during a vehicle duty cycle will not exceed 80 percent of the rated battery capacity. This will
be necessary to maintain battery life over many discharge
cycles. The requisite replacement life of the battery to
achieve market acceptability will depend on the resale
market for the vehicle and the salvage values of the battery sets, but in no case should the battery life fall below the expected vehicle ownership period for the original
purchaser. This has been estimated at 5 yr or 30,000 miles
for the shopper/commuter vehicle, and 10 yr or 100,000
miles for the family car.

Safety and Environment

Comprehensive safety and environmental standards for vehicles have been set by National Highway Traffic Safety
Administration and Environmental Protection Agency regulations. Electric vehicles and their storage systems will
no doubt be required to meet these same standards. These
should impose no insuperable difficulties, although some
of the more advanced batteries that operate at high temperatures may need special attention under crash conditions.
(Special safety considerations for flywheels in hybrid
electric installations will be discussed under the parameters for hybrid flywheel vehicles.)

Cost

Acceptable costs for electric vehicles will be entirely
dictated by the comparative costs of alternate vehicles,
presumably conventional automobiles with internal combustion engines. As discussed earlier, although the first
costs of fleets of electric vehicles are almost twice those
of comparable gasoline vehicles, fuel and maintenance costs
are sufficiently low to give comparable life-cycle costs--
a primary concern for the fleet owner. The personal electric vehicle, which represents the major market opportunity,
will probably not be attractive to individual buyers unless
the first cost is comparable to that of gasoline-powered
vehicles, since these buyers are more concerned with first
costs than life-cycle costs.

In the absence of detailed configuration designs, the
subportion of the electric vehicle costs that can be assigned to the battery storage system cannot be determined.
However, unit energy costs for the advanced battery systems
will probably have to be far less than the $30-$40 per
kilowatt-hour for lead-acid batteries that are made by the
millions each year.

Hybrid Flywheel Vehicles

The technical characteristics for flywheels operated in
conjunction with gasoline engines can only be roughly
approximated because of the many possible designs.

Size

The flywheel and transmission linkages must be capable of
installation within the vehicle envelope without intrusion
on the passenger compartment. The weight, including
protective shields, must be low enough to maintain the
performance of a gasoline-powered vehicle.

Storage Capacities

Because of the short duration of the power extractions and
the presence of the gasoline engine for recharging, the
stored energy capacity of the flywheel need only be modest.
Several different unpublished design studies reported to
the committee by the U.S. Department of Transportation

have indicated that 0.2-0.25 KWh of stored energy will suffice. This would correspond to the kinetic energy of a 3,000 lb vehicle at 55 mph.

Charge/Discharge

In a typical driving cycle, high power is needed for only a few seconds, with relatively long intervening periods. Flywheel discharge rates of 50-60 KW would provide about 15 s of full power. Much lower charge rates would be sufficient for normal recharge. However, if the flywheel is used for regenerative braking, maximum charging rates comparable to the maximum discharge rates will be needed.

Lifetime

Preferably, the flywheel should last as long as the gasoline engine with which it is coupled.

Safety and Environment

The flywheel will probably not generate any environmental problems. The vehicle will have to meet applicable crash safety standards relating to the safety of vehicle occupants. In addition, at least three special safety considerations need to be examined and solved for the flywheel with respect to community safety.

- Flywheel breakup from operational failures must be contained, both for the protection of the vehicle occupants and for the surrounding community.
- In the event of an abrupt flywheel seizure, means must be found to prevent the translation of the flywheel kinetic energy into vehicular rotation that would be hazardous to other vehicles.
- The flywheel must be prevented from breaking free of the vehicle in an accident, while still possessing high kinetic energy.

Cost

The same general economic parameters will probably apply for hybrid storage vehicles as for electric vehicles.

Life-cycle costs will be important to fleet operators, but individual users will probably emphasize first costs. Accordingly, sizeable penetrations of the personal vehicle market will probably require that the hybrid power system not exceed the costs of a competitive, all-gasoline-powered vehicle.

CONCLUSIONS

The market for vehicles powered with externally generated, stored energy as contrasted with an on-board internal combusion engine using stored fuel will depend heavily upon the vehicle use. For specialized commercial applications, particularly those with short-range, multistop, defined routes, electric vehicles with low maintenance and energy costs may produce life-cycle cost savings that will attract fleet owners despite comparatively higher capital costs. Historically, however, the individual car owner has paid little or no attention to life-cycle cost and has largely been influenced by performance, convenience, and first-cost considerations. A sizeable penetration of the personal car market will probably await the incorporation of vehicle characteristics, including cost, that are competitive with those of gasoline-powered cars.

To be competitive, electric vehicles will require battery systems with much higher energy and power densities than are currently available. The advanced batteries will, however, also need to cost considerably less than current lead-acid batteries if competitive first-cost criteria are to be met for the electric vehicle.

The committee concludes that the development of a competitive electric family vehicle (i.e., 4-6-passenger capacity, 200-mile rante, 55-mph speed capability, and at a competitive cost) will require at least 25 yr of research and development. An urban shopper commuter vehicle of limited (50-mile) range might be technologically feasible in about 10 yr.

In contrast to vehicles using all stored energy, combining storage with a conventional internal combustion engine might enable the earlier development of an acceptable alternative to the all-gasoline vehicle. A flywheel, for example, coupled with a small gasoline engine, might preserve the range and flexibility of the gasoline-powered vehicle while markedly improving the mileage and reducing emissions. The flywheel concepts need further development,

and the special safety problems associated with the flywheel installation have not yet been adequately explored,
but such hybrid systems warrant it. As in the case of all-
electric vehicles, the comparative costs of hybrid flywheel
vehicles (as yet unknown) will have a major influence on
the market acceptance of such vehicles.

6 SOLAR-ELECTRIC SYSTEMS

The four preceding chapters have considered energy storage for user sectors of the present economy. This chapter will consider the storage aspects of a technology that is not yet operational--solar-electric power generation. Broadly speaking, solar-electric generation encompasses any technology used to convert energy from the sun into electricity. This chapter, however, will deal with only those technologies where the electrical output is subject to a regular or frequent cessation due to solar input variations. These include solar-thermal-electric, photovoltaic, and wind-power systems.

The need to address storage in the context of specific energy sources and systems is particularly evident in the case of solar-electric systems. In conventional utility systems, storage is only an optional consideration for improving the match between fixed generating capacities and variable electrical loads. Solar-electric systems have an inverse problem. The variable generating capacity must be matched to a load that is either constant or variable on a different cycle. This requires, for all practical purposes, a buffering storage capacity between the solar power input and the load, or a backup generating capacity. For example, Figure 4 illustrates a hypothetical power-generation-load relation where the load primarily comes from daytime air conditioning needs.

The need for storage is absolute if the generating system is free-standing, but may be minimal if the solar-electric generator is a small input component (4-8 percent) to a large, interconnected utility complex.

The specifications and criteria for the storage systems will vary according to the specific generating concept, the operating mode (i.e., base-, intermediate-, or peak-load services), the location and the corresponding regional

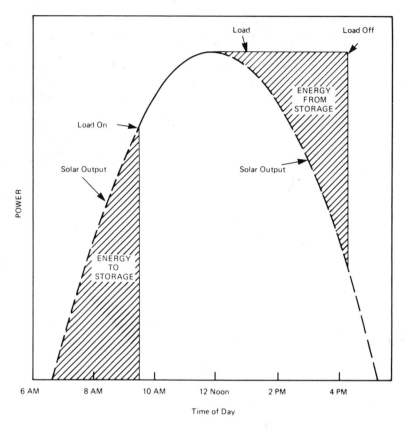

FIGURE 4 Hypothetical match of solar power system
and air conditioning load.

meteorology, and the degree of integration with a larger
utility system. All three solar electric systems are amen-
able to conventional utility storage approaches downstream
of the generator. In addition, flywheel storage could be
employed on a wind-power system between the windmill and
the generator, and thermal storage could be placed ahead
of the turbogenerators in a thermal-electric system.

In addition to the storage needed to match the daily
solar input and the load cycles, in some instances very
short-term storage may be needed to accommodate momentary
supply interruptions, such as the passing of a cloud. The
several systems will vary in their response to such interruptions. Photovoltaic systems will respond instantaneously.
Windmills will respond to a change in wind speed very

quickly, but they do have some mechanical inertia. The
thermal systems will have somewhat greater intrinsic storage capacities in the form of the working fluid and its
container and thus will have the lowest transient response of the three systems.

Specific storage criteria are discussed separately for
the three solar-electric-generating concepts in the following sections. The technical criteria are generalized for
a range of conditions in the country, although some of the
basic analyses were conducted for a specific region. Common
economic criteria are discussed for all three concepts.

TECHNICAL CRITERIA

Solar-Thermal-Electric Plants

The storage criteria include requirements for both on-site
and utility generation of base-load, intermediate-load, and
peak-load electricity. The criteria are generalized so as
to be as regionally independent as possible, although the
studies on which the criteria are based were done for a
specific region.*

Storage Capacities

The amount of storage needed will depend on the load that
the solar plant is expected to supply. Base-load plants
operating for as long as 7,500 h per year will require storage for 15-18 h to supply power at standard reliability criteria (a loss of load probability of 1 day in 10 yr).
Intermediate-load plants operating in the range of 1,000-
5,000 h per year will require storage durations of from 3
to 12 h. Storage of 1-6 h will suffice for peaking-load
plants operating for less than 1,000 h per year. The corresponding storage capacities range from 3,000-3,600 MWh,
300-1,200 MWh, and 30-180 MWh for base-load, intermediate,
and peaking plants.

Small, on-site solar plants operated on a local basis,
either independently of utilities or drawing on the utilities for backup electricity, may require up to 18 h of storage with corresponding capacities of 0.1-100 MWh.

---

*The data presented, as well as most of the information on
solar-thermal-electric power plants, were obtained from
reference 10.

Charge/Discharge Ratios

The charge/discharge ratio is defined here as the rate of
energy supply into the storage facility divided by the rate
of energy withdrawal, or the ratio of power in to power
out. Ratios of 2.2, 2.7, and 3.3 can be computed for base-
load, intermediate, and peaking plants, respectively. The
corresponding ratios for on-site solar plants will range
from 0.5 to 2.0. In these computations, a storage efficiency of 0.85 was assumed, with corresponding storage
losses of 0.1 percent per hour. This level of efficiency
was assumed to be independent of the type of solar-collector system (central receiver, parabolic cylinder, or parab-
oloidal dish).

System Lifetime

The storage-system component of a solar plant should have
a lifetime as long as the basic plant--normally 30 yr.
Systems with shorter lifetimes may be acceptable if life-
cycle cost analyses establish their economic viability.

Physical Limitations

Solar plants will require large land areas for their collector subsystems. The size of the area required will
vary directly with the plant capacity, type of load, type
of collector, and storage magnitude and duration. Typically, a 100-MW base-load plant with 12 h of storage might
require collection areas of 1.5 Km^2 with a central receiver
type collector, 1.8 Km^2 with a paraboloid dish, and 2-3 Km^2
with a parabolic cylinder system, depending on the orien-
tation of the cylinders. In comparison, the areas needed
for energy storage should present no special problems or
limitations.

Other Criteria

The safety and environmental criteria for the storage
components of solar plants will, in general, be the same
as those of the basic plant. Overall, solar plants will
most likely be sited in desert regions, where the availability of cooling water may be a problem. The lower
efficiency of solar plants compared to conventional plants

will accentuate the need for water. For example, the
annual cooling water requirements for a 1,000-MW base-load
solar plant using a conventional thermodynamic cycle and
operating at 80 percent load factor (7,000 h/yr) are estimated to be 20,000-25,000 acre-feet. The impact of the
storage system on these requirements should be examined on
a case-by-case basis.

The maximum accident postulated for a solar plant is
a rupture of a heat transfer line with a consequent spill-
age of the heat transfer and/or storage media. Safety
precautions should be developed for the storage system to
guard against hazards of contaminations for such leakages.

Photovoltaic Power Systems

The requirements and criteria for photovoltaic storage
subsystems can only be defined in a limited way at this
time. This definition will improve with time as system
studies yield more detailed results, as other subsystems,
circuits, and components become better defined, and as the
performance and costs become more certain. In the absence
of such information, the arbitrary division of photovoltaic power systems according to size and problem categories
corresponding to residential, commercial and small industrial, and utility-size systems is rational in that the
size of the photovoltaic component as a fraction of the
total interconnected power system makes a great difference
in the storage needs.

Residential Units

Photovoltaic systems for residences will be sized to provide all or part of the electrical needs of individual
residences or small apartment complexes. The committee
assumes that space and domestic water heating will be
supplied by thermal energy obtained directly from solar
or other sources rather than from a photovoltaic system
because of the low conversion efficiency. At least some
of the systems will probably employ combination thermal/
photovoltaic collectors that will provide both low-tempera-
ture heat and electric power. Further, at least some of
the systems will probably use heat pumps for summer cooling and auxiliary winter heating when the stored thermal
energy is not adequate to meet the demand. The following
section is concerned only with the electrical storage
requirements, not the energy stored in the form of heat.

Short-term storage for the smooth operation of the electrical system during brief interruptions of the sun (periods of 0.1 to 1.0 h) would have to go from an electricity acceptance system to a delivering system almost instantaneously. The amount of storage would need to be of the order of 0.5-10.0 KWh, capable of discharge at a 5-10 KW rate. Incorporating such short-term storage into the electrical system might be desirable even where a supplementary connection to a utility grid exists and storage would not otherwise be planned for.

Storage for supplying electricity during dark periods (12-24 h) could be diversified so that the energy could be stored in the form it will be used--thermal, electrical, or mechanical. Covering residential electricity, exclusive of heating and domestic hot water loads, will require a storage capacity of 8-25 KWh, capable of discharge rates of 5-10 KW. Providing heat and domestic hot water in a well-insulated home by an electrically driven heat pump would increase the required storage capacity by 12-50 KWh, with a charge rate of 5-20 KW, and a discharge rate of 1-4 KW. The heating requirement will be lower in the South but higher in some areas of the North. The winter heating load in the South will be replaced by a larger summer cooling load, which, however, will be more coincident with the available sunlight and require less storage.

The upper limits of the ranges given apply for large single-family residences with 24-h storage capability. Apartment units would require about half this storage capacity and half the discharge rates per unit. The lower limits of the ranges apply to shorter storage times.

Storage for long periods of low insolation will be very sensitive to geographical location because of the large differences in available sunlight. In the Northeast, 10-day periods of cloudiness occur occasionally, during which stored energy may have to be used if other forms of auxiliary energy are not available or practical. In such a case, capacities up to 250 KWh for large homes without electric heating and up to 750 KWh with heat-pump operations may be required. The discharge rates will not differ from those during dark periods, however.

Commercial and Industrial Units (Moderate Size)

Installations ranging from slightly above residential consumption levels to several hundred megawatts of electric power may be candidates for total energy systems. In any

case, the systems will need to be planned in an integrated
fashion.

Short-term storage during temporary cloud shadows would
be needed. This could be coupled with load control to reduce the amount of storage capacity required, but the
quick-off and quick-on features of storage are essential.
If solar energy becomes a major power source, the utilities would not be able to handle the rapid, large power
fluctuations that mixed cloudy and sunny days would
produce.

The dark-period storage needs of industrial and commercial units will vary considerably. In both cases, the
primary energy for heating and cooling should be stored
thermally, with electrical storage intended for heat-pump
rather than resistance heating. High-temperature, industrial process heat may, however, require direct electrical heating.

The amount of storage needed by a commercial unit should
be quite predictable from the operation and the amount of
sunlight available. Generally, 4 h of storage at an
average load should be sufficient for commercial units.
An industrial plant operating on a one-shift basis would
have low storage needs. A three-shift operation would
need storage for 18 h at average load if the electrical
needs are to be completely supplied by photovoltaic systems.

The long-term storage needs of commercial and industrial
units will be similar to those of residences at the low
end of the size scale and to utilities at the high end.
Discharge rates will generally be equivalent to dark-
period operations, but charge rates may have to be several times as high as those of the dark-period systems. In
geographical areas with high insolation, the long-term
storage requirements will not have to be much greater than
those for the dark periods.

Utilities

The demand for short-term storage capable of quick response to partial cloudiness must be met by utilities.
Energy outflow must be available in a fraction of a cycle
for the utility, and it must continue until the sun again
shines or until other generation units can be brought on
the line. The need for other backup generation to the
photovoltaic system for short-term fluctuations would be
a serious operational limitation. For a 1,000-MW system,
at least 500 MWh of storage would be necessary to avoid
drastic load curtailments.

For utility purposes, dark-period and long-term storage would be intergrated with other operational requirements for storage.

Common Considerations

Some considerations for photovoltaic systems are common to residential, commercial, and industrial buildings as well as to utilities. Alternative load controls and utility grid connections would affect the storage needs. The optimum combination would be largely determined by local considerations. At one extreme, the utility might provide most of the storage. At the other, a free-standing photovoltaic system might have its own auxiliary power system. In the ideal case, the auxiliary energy converter would be a part of the same system that provides the dark-period storage.

As an example, an electrolysis-cell/fuel-cell system with hydrogen and oxygen tanks could be the storage system. Auxiliary energy could be provided from additional hydrogen and oxygen delivered periodically by tank truck, stored locally in enlarged tanks, and converted by the same fuel cell. Such an auxiliary system is characterized by a small average load and a large-peak/average-load ratio. Such characteristics are very unfavorable for fixed distribution grids, but they can be equalized by provision of local storage.

Because of the anticipated relatively high cost of energy generated by photovoltaic conversion, due to the expected high capital cost of the conversion subsystem, the storage output/input efficiency must be as high as possible for economic reasons.

The storage capacities are "useful" capacities rather than "nominal" capacities. Some storage devices, such as batteries, cannot be discharged below a lower fraction of the nominal capacity nor be charged above a higher fraction without reducing the device's lifetime. The "useful" capacity is the energy output actually available from swinging the storage device between these two fractions and after deducting storage losses.

A useful life for the storage subsystem of at least 20 yr, or 7,500 deep-charge/discharge cycles is desirable; attainment of at least half of these values is probably a necessity. Maintenance costs during the lifetime should be minimal. Periodic servicing should preferably be of the "do-it-yourself" type, such as cleaning filters or exchanging easily accessible and readily replaceable parts, which can be carried out by homeowners or relatively un-skilled maintenance personnel.

Volume and mass of the storage subsystem are of second-ary importance. High-temperature operation and moving parts (e.g., flywheels), although not desirable for resi-dential and commercial installations, may be tolerated since the storage systems are stationary and can be shield-ed and insulated.

The charge/discharge ratios required of the storage systems will probably not seriously constrain the selec-tion of the type of system, except in the case of short-term storage. In most instances, the peak charge rates will be higher than the peak discharge rates because the storage system has to accept the incoming solar energy up to the capacity of the photovoltaic converter as long as the storage is not filled.

Wind-Power Systems

Wind-driven generators (WDG's) are a form of solar energy recovery that is not directly dependent on local insola-tion. Single units ranging from 50 KW to 3 MW are being considered. Mean wind speeds of from 12 to 24 mph will suffice for their operation. Because of the low unit sizes and the variability of winds, the WDG's are being consid-ered primarily as peak-load machines. The discussion of storage criteria that follows is confined to utility owner-ship and operation of the WDG's and the storage system.

For operation as a peak-load generator, the WDG elec-trical output during off-peak hours will be used to charge a storage system. The WDG will feed the load directly during the peak-load periods, up to the limit of its capa-bility, and supplemental energy, as needed, will be drawn from the storage system.* The amount of required storage will depend on the speed and constancy of the wind rela-tive to the WDG design parameters.

Northeast Utilities (of Connecticut) has been studying the use of stored wind power to meet specified peaking electrical loads. In lieu of generalized assumptions about wind conditions, Northeast Utilities has acquired wind data four times an hour at four altitudes over

---

*A fuller description of the basic WDG's and their system interconnections is given in reference 11.

several potential WDG locations. The data have been used
to determine the generating capacity and storage requirements to meet a 5-MW load during weekday hours of 9-11 a.m.
and 1-3 p.m. The storage cirteria that follow were developed from the unpublished results of that study.

Storage Capacity

The amount of storage capacity required to meet specified
loads is extremely dependent on the site. Both the average speed and the constancy of the wind are major factors.
For example, the Northeast Utilities study showed that for
a WDG on a Long Island Sound shoreline site having an average wind speed of 12 mph at a 150-ft altitude, the specified 5-MW, 4-h peak load could be met with 5 WDG's of 1 MW
each and 5 h of full-load storage capacity. By contrast,
an inland site with an average wind speed of 9.3 mph at a
150-ft altitude would require the installation of 10 MW
of WDG's and 33 h of storage capacity to meet the same
load.

The storage capacity will also be strongly dependent on
the required system reliability. The reliability is defined as the percentage of the annual predicated load that
can be met during peak-load periods from the stored energy
plus the direct WDG output. If the storage is depleted at
a time when the winds are insufficient to meet the peak
load, an "outage" occurs. The assumed 5-MW, 4-h peak load
in the Northeast Utilities study could be met with a 65
percent reliability using 5 MW of WDG's that might require
as little as 2 h of storage (or 10 MWh). An 80 percent
reliability could require 5 h (25 MWh) of storage.

Other WDG/storage combinations having the same energy
storage capacity might also meet the reliability criterion,
depending on the site. The study indicated that 80 percent
reliability could be met in one case with only 3 h of storage (coupled with 8 MW of WDG's). In another case, a system of only 4 MW of WDG's and 6 h of storage would also
meet the criterion.

Charge/Discharge

The charging rate of the system should be equal to the
combined capacities of the WDG's so that the full off-peak
capacity of the WDG system can be used to charge the storage system. The discharge capacity from storage should be

equal to the peak-load power rate to cover the possibility
that the on-peak contribution from the WDG's might be
temporarily zero during wind lulls.

The computed storage capacities in the Northeast Utilities study are based upon the assumption that the storage
device is able to exactly follow changes in either charge
or discharge rates. The operational desirability of this
is, of course, coupled to the moment-to-moment variability of the WDG output. Some storage devices are essentially on-off devices and cannot operate efficiently at
variable charge and/or discharge rates. Pumped-hydro
storage units, for example, may be unacceptable for many
wind-power (or other solar power) applications without
further development. Batteries, on the other hand, will
probably be able to adjust charge and discharge rates more
successfully. In fact, batteries improve their efficiency
at below design charge and discharge rates. The storage
options intended for development to meet solar-electric
system uses should be carefully evaluated for flexibility.

Other Criteria

The storage devices used with wind-power systems need not
be physically contiguous to the WDG's, but may be tied to
them through the electrical grid. For this reason, no
unique safety, environmental, or physical criteria are
evident for wind-power storage uses--the criteria for
general utility storage systems will apply.

ECONOMIC CRITERIA

The storage cost limits to make a solar-electric system
commercially attractive are difficult to isolate from the
overall solar plant cost limits. In order for the cost of
electricity from the complete solar plant to break even
with the cost of electricity from a competing generating
option, it can be shown that:

$$K_s = \frac{\frac{r_c K_c + h_c (O_c + 0.003413 \cdot f_c) - h_s O_s}{e_c}}{r_s},$$

83

where

K = Capital investment, \$/KW
r = Annual carrying charge rate
h = Duration of annual operation, h
O = O&M costs, \$/KWh
f = Fuel costs, \$/10^6 Btu
e = Fuel conversion efficiency, %

and, subscripts

s = Solar plant
c = Competing plant.

If, for purposes of simplification, annual hours of operation, annual carrying charge rates, and O&M cost rates are assumed to be the same for the solar and competing plants, then the equation reduces to:

$$K_s = K_c + \frac{0.003413 \cdot f_c \cdot h}{e_c \cdot r}.$$

This equation shows that for break-even electricity costs, the added solar plant investment (as compared with the competing plant) is dependent upon the annual fuel savings.

For example, a solar-electric plant competing with a coal-fired base-load plant on a 7,000 h/yr duty cycle with coal costs at \$1/million Btu, a conversion efficiency of 0.36, and an annual carrying charge rate of 0.15 could have an extra investment as high as \$440 per average kilowatt. Similarly, if the competing fuel cost were as high as \$2.5/million Btu (corresponding to oil at \$15/bbl), the added solar plant investment could be as much as \$1,100 per average kilowatt more than the competing plant for the same base-load cycle. If, on the other hand, the solar plant were planned for peaking duty of only 1,000 h/yr, then at the higher competing fuel costs and conversion efficiency of 0.28 (typical of an oil-fired gas turbine peaking plant), the break-even solar plant investment increment would drop to only \$200 per average kilowatt.

In the above examples, added solar plant investment would have to include the storage system needed to attain the postulated annual operations. The permissible storage costs cannot be evaluated as independent plant elements.

CONCLUSIONS

Because of the fluctuating nature of the energy input, solar-electric power systems will, in general, need storage systems to match demand loads. Depending on the type, scale, and interconnections of the solar plant, different types of storage devices may be needed--quick response units to accommodate momentary input interruptions and large capacity units to provide power during night or extended low-insolation periods.

Although overall storage requirements may be similar to those of conventional utilities, the variable and, to some degree, unpredictable input power levels may lead to different storage preferences. In particular, a higher value will probably be placed on storage devices capable of handling variable charge and discharge rates as contrasted with fixed-capacity electromechanical devices. A higher value will also likely be placed on devices that can undergo frequent cycling from charge to discharge modes.

Isolated cost targets for integral storage elements of total solar-electric power systems cannot be specified. Since, however, overall solar plant costs, as compared to conventional generating systems, may be the greatest obstacle to their widespread deployment of such systems, the storage component costs must be as low as possible.

7 FUSION REACTORS

There are many conceptual approaches to the production of fusion reactors, often differing in both their physics and their technologies. Since scientific feasibility has not been demonstrated for any of the approaches, it is too early to predict whether fusion reactors will become practical, commercial components of the national electrical supply system. However, conceptual, full-scale fusion reactor designs have emerged in the past 3-5 yr. Though these are only the first attempts at designs that will be iterated for the next few decades, they indicate a need for energy storage and transfer systems of a size and transfer rate exceeding any existing today. This section briefly outlines the storage requirements to emphasize their potential importance.

In the absence of a feasibility demonstration for the reactors, it is not known which of the storage systems will be needed. To further compound matters, the embryonic nature of the reactor designs does not permit an exclusive listing of storage systems, nor do their requirements have a precise measure. Despite such uncertainties, a great deal is known in a general sense, and general system requirements can be listed for the designs at hand. In the following discussions, the ranges given for the requirements indicate the uncertainties of the specific design concepts, variations between similar devices designed by different design teams, or they simply reflect the lack of a specific design and represent educated guesses. Such is the current nature of understanding of fusion reactors.

REACTOR DESIGNS

At present, magnetic confinement concepts center on the tokamak, mirror, and theta pinch devices. Inertial confinement devices include laser or relativistic electron beam (E-beam) impingement on pellet targets. Many other approaches are being funded for study in the United States and elsewhere in the world. The most promising of these include linear systems heated by lasers or E-beams, high-power density tokamaks, steady state toroidal reactors, toroidal Z pinches, and several more speculative systems based on imploding metal liners.

The following requirements are taken from the few extant designs for the main approaches. All of the studies have been for fusion plants with electrical outputs in the range of 500-3,000 MW. The mirrors operate at a steady state; the tokamaks are quasi-steady, having 2-10-min burns and a greater than 50 percent duty cycle; and the theta pinches have a 0.1-s burn every 10 s. Liners, lasers, and E-beam systems would have even shorter pulses with higher repetition rates. Since most of these devices handle or store energy in amounts comparable to or larger than their output in one pulse, or have power-conditioning systems whose power is comparable with the plant output, the characteristics of the burn cycle have a large influence on the above-mentioned systems.

ENERGY STORAGE REQUIREMENTS

Six kinds of energy-handling systems are characteristic of fusion systems. They are:

1. Large superconducting magnets for plasma confinement, 10-250 GJ (3-70 MWh), steady state;
2. pulsed magnetic field systems, 1-50 GJ (0.3-14.0 MWh), 50 µs to 10 ms;
3. inertial energy systems, 1-50 GJ (0.3-14.0 MWh), 10 ms to 5 s;
4. capacitor systems, 1-10 MJ (0.3-3.0 KWh), 0.1 to 5.0 µs;
5. power-conditioning systems, 0.5-2.5 GW, 1-10 s; and
6. energy-leveling systems, 100-500 GJ (30-140 MWh).

The estimated ranges for these systems, where applicable, follow for several of the principal fusion concepts.

Tokamaks

1. 150-250-GJ (40-70-MWh) toroidal superconducting magnet set, steady state;
2. 3-10-GJ (0.8-3.0-MWh) pulsed ohmic heating and equilibrium field coils (superconducting);
3. 300-500-MW auxiliary heating (neutral beams or R.F.) 3-10-s duration, 200 Kv; and,
4. 100-500-GJ (30-140 MWh) load leveling for low-duty-cycle (less than 50 percent) operation.

The duration of the burn pulse has a significant influence on hte pulse systems (2 and 3) since their efficiencies are determined by the requirement of a net energy balance. The rotating machinery and switches that transfer energy to and from the pulsed coils, the coils themselves, and the auxiliary heating devices are beyond the present state of technology.

Energy leveling will be necessary if the reactor is off longer than the thermal time constant of the blanket. Also, if plant output is variable during the burn, either a control system or an energy-leveling system may be required.

Theta Pinches

1. 60-GJ (17-MWh) toroidal (room temperature) field coil, 30-ms transfer, 100-ms hold.

For this system, inertial storage using superconducting field winding homopolar generator units of approximately 1 GJ 9278 KWh), each designed for fast transfer, have been advanced. An efficiency of 95 percent for a complete delivery and recovery cycle is needed for a favorable energy balance.

Mirrors

1. 25-50-GJ (7-14-MWh) Yin-Yang mirror coils, steady state, 150 kilogauss superconducting;
2. 2,400-MW neutral beam heating, steady state, 200 Kv; and,
3. 2,400-MW particle decelerators (direct converters), steady state.

The magnets must operate at the highest possible field. The neutral beams are negative ion sources to provide high neutralization efficiency. An overall neutral beam injector efficiency of greater than 80 percent is required. The direct converter must decelerate approximately 50-200 Kv ions at greater than a 70 percent efficiency to allow an energy balance. The power output is approximately 900 MW.

Liners

1. 5-10-GJ pulsed liner implosion energy, 50-100-μs delivery.

The metallic liner implodes a trapped magnetic field to approximately 1-2 megagauss, allowing a short linear system. Reexpansion of the plasma against the magnetic field returns energy to the source, which must be capable of recovering and storing it at high overall efficiencies.

Lasers

1. 1-10-MJ (0.3-3 KWh) laser supply, 5-500-μs delivery.

For gas lasers, short (5-10 μs) delivery times are required. Glass laser flash lamps need approximately 1/2-ms (possibly inductive) supplies.

USER CONSIDERATIONS

Utility system economics, reliability, and control considerations will also constrain the storage systems for fusion reactors. Studies of the fusion concepts have projected costs of $3,000-$5,000 per kilowatt output. Obviously, these costs will have to be reduced if the fusion plants are to compete with more conventional forms of electricity generation. Although the balance-of-plant rather than nuclear island costs dominate, the storage costs for the reactor will be significant. Further, the normal contingency, engineering, capitalization, and associated plant costs amplify the importance of reducing the basic storage costs.

More expensive systems might be required in certain instances to achieve an acceptable level of component reliability. In general, about 80 percent generating availability would be necessary for the fusion plants, with about 20 percent scheduled down time for maintenance and forced outage. A minimum 20-yr lifetime and a 1-2 yr mean time-to-failure for replaceable parts would seem reasonable.

The storage-system requirements are also affected by control of the overall plant output. Feedback control of neutral beam injectors, magnetic fields, or other fusion energy subsystems might be required for a constant output during the burn. Repeatability from pulse to pulse might require these same systems, or, if there is a significant variation (as in present physics experiments), some large load-leveling storage system such as superconducting magnet storage or thermal storage in the blanket could be required. Finally, if no natural parameters, such as the pulse-repetition frequency, are available for varying the plant output, the above load-leveling schemes might also be needed for that purpose.

CONCLUSIONS

Some energy storage systems that will be used in the operation of a fusion power plant will handle large amounts of energy compared to the plant output. They will constitute a significant fraction of the plant costs and can adversely affect the energy balance in the reactor. Much of the anticipated technology is beyond the state of the art, and the system requirements are, as yet, poorly defined. From the general features of the systems and the range of values characterizing them, the importance of specifying the problems more clearly and finding appropriate technical solutions is obvious.

Because of the integral and specialized character of the storage systems that will be required for the basic operation of the fusion plants, their solution will often best be handled by the fusion plant research and development teams rather than by more general energy storage specialists. However, there are instances where some independent engineering of storage systems is needed to better bound the costs and select technologies that are mutually compatible with the requirements of the fusion process and the total system operation. As fusion plant operating characteristics become more clearly defined, furthermore, there may be a need for buffering storage between the plant and the load of a more general nature.

8 EXPLORATORY R&D

The preceding chapters have presented the technical and
economic performance criteria that, in the judgment of the
committee, must be met by advanced energy storage systems
if they are to have reasonable prospects of marketplace
acceptance. An expectation that the performance of storage systems will meet or exceed these criteria should,
properly, be a major consideration in deciding whether,
and to what extent, advanced development and demonstration
R&D activities are warranted.

Exploratory research, on the other hand, conducted to
identify new concepts, principles, or physical parameters
to improve storage devices and systems should not be
evaluated by the same criteria.

The committee was impressed by the many potential areas
for storage applications and by the possibilities for high
returns in energy savings, cost reductions, and conservation of scarce fuels. At the same time, although the
candidate technologies were not formally evaluated, the
committee was concerned by the rather marginal ideas for
fully appropriate storage systems. Therefore, a vigorous
exploratory research program should be pursued to search
for new ideas. The payoff potential could be enormous.

In this search for new ideas, the criteria for research
sponsorship should not include the requirement for identification of ultimate commercial feasibility. Frequently
the initial concept of the ultimate application of a new
idea is not subsequently borne out. There may be unconsidered problems that leave the research without practical
merit for the anticipated application. On the other hand,
there may also be unconsidered opportunities that magnify
the apparent initial value of the research. Premature
screening of proposed exploratory research using ultimate
application criteria is, therefore, inappropriate.

The committee does not advocate the conduct of pure, undirected research under the rubric of exploratory energy storage research. The sponsored research should be directed at identifiable storage problems. Proposals for such research should be evaluated solely in terms of the qualifications of the investigator and the potential contribution of the research to the solution of unresolved problems or to the identification of new concepts of energy storage.

9 CONCLUDING REMARKS

Until recently, energy was too cheap for most consumers to have an economic incentive to make capital investments to reduce energy consumption and costs. Energy prices have now risen to a level, however, that has greatly increased the worth of such investments. Furthermore, there is a growing prospect that world petroleum resources will, perhaps within a few decades, be inadequate to satisfy the growing demand for oil. This could drive energy prices to even higher levels and would accentuate the need to develop fuel-substitution techniques for many current oil and natural gas uses. The exploration of cost-effective techniques should be given a high priority. Even if fuel substitution results in an absolute increase in energy consumption from more abundant fuels, the principle of conserving scarce fuels needs emphasis.

Conceptually energy storage systems have an enormous potential for facilitating cost-effective, large-scale fuel substitutions. In some applications, they could also reduce absolute energy consumption. The realization of this potential will depend upon the achievement, through research and development, of proper technological and economic operating characteristics.

This study has developed criteria to determine whether prospective advanced energy storage systems will have the performance characteristics that will make them attractive to ultimate users and, therefore, worth pursuing through advanced development and demonstration stages. In some areas the criteria are specific and quantitative. In others, and particularly for economic parameters, the criteria have only been expressed in qualitative terms because of the diversity of potential applications (as in industrial and residential/commercial areas), the complex trade-off factors (utilities), or the present technological uncertainty of the applications (solar-electric and fusion reactor areas).

The criteria are intended as guidelines for evaluating the potential merits of R&D projects that might find future applications. These merits are not absolute, but will depend upon the conditions that will exist in the period following the completion of the R&D. Since the relationships between the availability of energy, its costs, and capital investments may change substantially over the next several decades, care should be taken not to apply too narrow a range of forecasts in making the evaluations, particularly for the economics of storage.

Care must also be taken to evaluate storage systems in terms that recognize their full potential impact. Although the committee did not formally evaluate specific storage technologies, an analysis of the opportunities indicates that some systems might satisfy several areas of application. For example, the storage of electrical energy for subsequent reuse as electricity is an important opportunity in all of the energy-use sectors. The value of advanced storage battery R&D would, therefore, be enhanced by the many potential applications. Similarly, electricity stored in a chemical form, as in hydrogen produced from the electrolysis of water, might have very widespread use, not only for the subsequent regeneration of electricity, but also for use as a direct fuel or as a basic chemical resource. Such versatility should be reflected in the evaluation of specific systems.

The criteria can be useful for evaluating advanced technology development and system demonstration activities to determine if they warrant R&D support. The criteria are inappropriate for evaluating exploratory research proposals. Such proposals should be evaluated solely in terms of the qualifications of the investigator and the potential contributions of the research to the solution of unresolved problems or to the identification of new concepts of energy storage. Screening of exploratory research using ultimate application criteria is premature and, perhaps, counterproductive.

As a final note, it is emphasized that energy storage is a complex topic that cannot be properly evaluated without a detailed understanding of both supply and end-use considerations. Storage benefits will largely arise from two impacts on the supply-demand structure: Storage will allow a more effective use of capital equipment by improving capacity factors and will permit cost-effective substitution of plentiful for scarce fuels. In general, a coordinated set of actions will have to be taken in several sectors of the energy system for the maximum potential benefits to be realized.

As a case-in-point, the growing use of electrical heating for residential and commercial buildings in place of oil or gas could exacerbate the peak-demand loads on electric utilities. From their viewpoint, the utilities might justify limited storage investments to improve their capacity factors. From a broader viewpoint, however, considering the electricity as only a link in the functional space-heating chain, storage investments at the point of end use of the thermal energy might be more prudent. In the United States, such investment decisions are outside the control of the utilities, and rate structures do not provide appropriate signals for optimal investment at present.

To encourage the implementation of residential and commercial storage systems may require several regulatory actions. Rate schedules will need to be adopted that will make the storage of off-peak energy economically attractive. Other actions may be necessary to ensure that the utilities plan a proper mix of coal-fired or nuclear-generating capacity to satisfy the increased electrical heating demand. It may also be desirable to adopt policies that allow utilities to lease storage equipment to consumers, thereby enabling them to incorporate end-use storage into their systems planning.

REFERENCES

1. Public Service Electric and Gas Company. *An Assessment of Energy Storage Systems Suitable for Use by Electric Utilities.* Prepared for the Energy Research and Development Administration (ERDA) and the Electric Power Research Institute (EPRI). (To be published in September 1976 by ERDA, Washington, D.C.)

2. F. R. Kalhammer and T. R. Schneider. "Energy Storage." *Annual Review of Energy.* Volume 1, 1976. Annual Reviews, Inc., Palo Alto, Calif.

3. Stanford Research Institute, Menlo Park, California. *Patterns of Energy Consumption in the United States.* January 1972.

4. A. L. Berlad, F. J. Salzano, and J. Batey. *Energy Management in Buildings: An Analysis and an Integrated Approach.* Brookhaven National Laboratory, Brookhaven, N.Y. BNL 20572, July 1975.

5. C. Braun. *Impact Analysis for Enthalpy Management in Building Systems.* Brookhaven National Laboratory, Brookhaven, N.Y. BNL 20524, September 1975.

6. Federal Energy Administration, Washington, D.C. *Project Independence and Energy Conservation: Transportation Sectors.* November 1974.

7. Federal Power Commission, Washington, D.C. *Development of Electrically Powered Vehicles.* February 19, 1967.

8. D. C. Sheridan, J. J. Bush, and W. R. Kuziak, Jr. *A Study of the Energy Utilization of Gasoline and Battery-Electric Powered Special Purpose Vehicles.* Society of Automotive Engineers, Warrendale, Pennsylvania. SAE Preprint 760119, February 1976.

9. Rockwell International. *Economic and Technical Feasibility Study for Energy Storage Flywheels.* Prepared for the Energy Research and Development Administration, Washington, D.C. ERDA 76-65, UC-94B, December 1975.

10. Aerospace Corporation, Los Angeles, Calif. *Solar Thermal Conversion Mission Analysis*. Prepared for the National Science Foundation/RANN Program, Report No. ATR-74(7417-05)-1, Washington, D.C. January 15, 1976.

11. G. E. Jorgenson, M. Lotcker, R. C. Meier, and D. Brierby. *Design, Economic and System Considerations of Large Wind-Driven Generators*. Submitted for presentation at the IEEE Power Engineering Society Winter Meeting, Institute of Electrical and Electronics Engineers, New York. January 1976.

UTILITIES PANEL

Chairman: Heinz G. Pfeiffer, Committee

Members: Harold B. Finger, Committee; Fritz R. Kalhammer,
Committee; Robert L. Bolger, Assistant Vice President,
Commonwealth Edison Company; Charles Goodman, Senior Re-
search Engineer, Southern Services, Inc.; Virginia T.
Sulzberger, Advanced Development Engineer, Public Service
Electric and Gas Company

Invited Participants: Deb Chatterji, General Electric
Company; Eldon Hall, General Electric Company; Raymond A.
Huse, Public Service Electric and Gas Company

RESIDENTIAL/COMMERCIAL PANEL

Chairman: Harold B. Finger, Committee

Members: Kenneth C. Hoffman, Committee; William E. Siri,
Committee; David C. White, Committee; Joseph H. Newman,
Senior Vice President, Tishman Research Corporation; Robert
Romancheck, Supervisor, Research and Technical Services,
Pennsylvania Power and Light Company

Invited Participant: Eldon Hall, General Electric Company

INDUSTRIAL PANEL

Chairman: Gerald L. Decker, Committee

Members: Fritz R. Kalhammer, Committee; J. R. Ferguson,
Jr., Executive Vice President, Engineering and Research,
U.S. Steel Corporation; Jack Hartung, Vice President,
Purchasing, St. Regis Paper Company; Edward W. Nicholson,
Planning Manager, Energy Systems, Exxon Enterprises; Neal
Richardson, Energy Research and Development Office, TRW
Systems and Energy

Consultant: Bruce W. Wilkinson, Associate Professor,
Chemical Engineering Department, Michigan State University

TRANSPORTATION PANEL

Chairman: Ronald Smelt, Committee

Members: Paul Chenea, Committee; Edward E. David, Jr.,
Committee; Robert A. Huggins, Committee; R. Eugene Goodson,
Director, Institute for Interdisciplinary Engineering
Studies, Purdue University; Lloyd J. Money, Assistant
Director for Systems Development, Department of Transporta-
tion

Invited Participants: Robert H. Cannon, Jr., California
Institute of Technology; Donn P. Crane, U.S. Postal
Service; David L. Douglas, Gould Incorporated; Serge Gratch,
Ford Motor Company

SOLAR-ELECTRIC PANEL

Chairman: Kenneth C. Hoffman, Committee

Members: Edward E. David, Jr., Committee; Heinz G. Pfeiffer
Committee; Ronald Smelt, Committee; Michael Lotker, Senior
Scientist, Northeast Utilities Service; Martin Wolf, Re-
search Associate Professor, University of Pennsylvania

Invited Participant: David L. Douglas, Gould Incorporated

FUSION POWER PANEL

Chairman: David C. White, Committee

Members: Robert A. Huggins, Committee; Lawrence M. Lidsky, Department of Nuclear Engineering, Massachusetts Institute of Technology; Keith Thomassen, CTR Division Office, Los Alamos Scientific Laboratories; Herbert H. Woodson, Chairman, Department of Electrical Engineering, University of Texas at Austin.

APPENDIX B:
UNITS AND CONVERSION FACTORS

The units used in this report, those customarily employed in each energy-use sector, are a mixture of metric and nonmetric units. Where customary units for power and energy are nonmetric, metric equivalents are also shown parenthetically. Other units, such as distance, speed, weight, and so on, are used in the text without conversion to their metric equivalents.

The basic metric units for power and energy are watts and joules, respectively, but at the levels encountered in many engineering applications power is expressed in kilowatts (KW) or megawatts (MW) and energy is expressed in kilowatt-hours (KWh) or megawatt-hours (MWh).

The following factors may be used for conversions:

Power
1 KW = 10^3 watts
1 MW = 10^6 watts

Energy
1 joule = 1 watt-second
1 megajoule (MJ) = 10^6 joules
1 MJ = 278 Wh
1 gigajoule (GJ) = 10^9 joules
1 GJ = 278 KWh
1 Btu = 1,055 joules
10^6 Btu = 293 KWh

Miscellaneous
1 pound = 0.454 kilogram
1 foot = 0.305 meter
1 mile = 1.61 kilometer
1 cubic foot = 0.0283 cubic meter
1 acre-foot = 1,234 cubic meters

100

APPENDIX C:
GLOSSARY

The following technical terms and abbreviations having
particular significance for energy storage have been used
in this report:

Add-on systems	Storage components that may be added optionally to basic space-conditioning systems
Base-load units	Utility generating equipment used to meet more or less continuous demand loads
BER	Building energy ratio. The space-conditioning load of a building expressed in thousands of $Btu/ft^2/yr$.
Capacity factor	The fraction of time that a specific class of utility generating equipment is in use
CCIF	Construction compound interest factor. A cost factor to account for the value of money during construction periods.
COP	Coefficient of performance. The ratio of the heat output to work input in a space-conditioning system.
Diurnal	Recurring daily
EPRI	Electric Power Research Institute
ERDA	U.S. Energy Research and Development Administration
Generation for load	The electric utility output to customers
Heat pump	A space-conditioning device that transfers heat mechanically to or from a reservoir, which is frequently the ambient conditions surrounding a building
Homopolar generator	A solid rotor machine without commutators or permanent magnets

Incremental valley energy costs	The cost of producing added power increments at or near minimum power demand levels
Insolation	The rate of delivery of direct solar energy per unit of horizontal surface
Installed reserves	Utility generating capacity that is in excess of maximum anticipated loads
Intermediate units	Utility generating equipment used to supply power for regularly recurring loads in excess of the continuous base load
Life-cycle costs	The accumulated capitalization, financing, maintenance, and operating costs of a device over its lifetime
Load factor	The ratio of the actual utility load to the rated capacity of the generating system
Mirror	A nuclear fusion device that confines the reaction within magnetic fields in such a way as to aid plasma confinement by particle reflection
O&M costs	Operating and maintenance costs
Peaking units	Utility generating equipment used for intermittent maximum loads
Photovoltaic	The process of directly converting sunlight to electricity through solar cells
PSEG	The Public Service Electric and Gas Company of New Jersey
Pumped-hydro	A utility storage system in which off-peak power is used to pump water to an elevated reservoir for later reconversion to electricity by passage through turbines at peak-demand periods
Retrofit	The subsequent installation of equipment after initial construction is completed
Solar thermal electric	The process of converting sunlight to electricity by heating an intermediate working fluid that drives a generator
T&D	Transmission and distribution
Theta pinch	A nuclear fusion device employing high plasma density, strong magnetic fields, and field stabilization feedback to product confinement. The device can be toroidal or possibly linear (if equipped with magnetic mirrors).